THE PRODUCTION OF
GRAPES
&
WINE
IN COOL CLIMATES

THE PRODUCTION OF
GRAPES
&
WINE
IN COOL CLIMATES

David Jackson & Danny Schuster

Daphne Brasell Associates Ltd
and
Gypsum Press

Reprinted in 2001 by:
Gypsum Press and
Daphne Brasell Associates Ltd
PO Box 12 214 Thorndon, Wellington
Aoteroa New Zealand
www.brasell.co.nz

ISBN 0-909049-17-3
First edition 1984 (Alister Taylor)
Second edition 1987 (Butterworths Horticultural Books)
Revised edition 1994 (Gypsum Press)
Reprinted 1997 (Lincoln University Press)

Printed by Astra Print Ltd, Wellington.

CONTENTS

Part II

Grape-Growing

Part III

Winemaking

Appendices

Tables

Maps

ACKNOWLEDGEMENTS

There are many people who have, in one way or another, assisted in the preparation of this book and it would be impossible to mention them all by name. We do, however, wish especially to thank our editor Michael ffolliott-Foster for his editorial skill and for his advice and encouragement. We thank the staff at the Educational Unit, Lincoln University and Graeme Steans from the Department of Horticulture for, respectively, photography and discussion.

Ross Beach designed the cover, Jan Hart, Marion Trought and Jaap Koster drew the diagrams, their undoubted skill is appreciated.

David Jackson
Danny Schuster

PREFACE

The aim of this book is to provide a background and guide to the production of grapes and wines, more especially in cooler regions. Its scope is considerable, covering the distribution and growing of grapes, the making of wine and its subsequent evaluation. Despite this, the material presented is not elementary and the detail will be quite adequate for a working understanding of all the processes described. To do this seemingly impossible task we have reduced or omitted a number of sections normally covered in detail in textbooks on grapes or wine; for example, in the section on viticulture we have devoted little space to varieties and techniques for warm climates. These are, in large part, covered in standard textbooks written predominately for such areas (references 2, 5, 6, 7). In concentrating on cooler climates we are acknowledging firstly, the increasing interest being shown in the more northern parts of the United States, in Canada, England, the cooler parts of Australia, Tasmania and the South Island of New Zealand and, secondly, the lack of written information, in English, for these areas.

Basic horticultural information has been greatly reduced, and this includes detailed comments on irrigation, nutrition, soil tillage, and the use of machinery such as sprayers, tractors, cultivators and mechanical harvesters. Important as these subjects are, changes from district to district and in design mean that information quickly becomes outdated and local advice can normally provide better guidance.

Diseases and pests also vary from district to district and would need too much space to be adequately covered in a book such as this — we only mention some of the main ones which cause special problems for the grape-grower. Growers can normally get excellent advice from local advisory officers and chemical distributors on current methods of control.

In the area of winemaking, basic principles are emphasised and we offer advice based on current commercial practices and recent research. We present detailed information where we think it is necessary, but do not enter into discussion on the many and varied types of crushers, presses and filters which are currently available. This is an ever-expanding field of development and like the machinery used for viticulture, varies with time and location.

We expect that this book will have special value in areas where grape-growing is not already a tradition, and where it is not always easy to get information on growing and winemaking techniques. In Part 1 we select, from around the world, a number of well-established and a few newer, cool-climate districts and, with the aid of tables of soils, geography and climate data, provide a background and guide to the major wines and grapes associated with these areas.

This conspectus is amplified in Parts II and III where we examine viticulture and oenology and consider, in detail, the requirements for successful grape culture and winemaking.

David Jackson
Danny Schuster
June, 1994

PART I

THE GEOGRAPHICAL DISTRIBUTION OF THE GRAPE

INTRODUCTION

Historical introduction

The original home of viticulture is Asia, the most likely area being the temperate, climatic regions of the Caucasus from where the wine grape, *Vitis vinifera* is said to originate. It is believed that the Assyrians and later the Pharoahs of Egypt enjoyed wine as long as five or six thousand years ago.

How the knowledge of grape-growing and winemaking spread further to Asia Minor and later to various islands in the Mediterranean can only be presumed, but it is known that the first merchant-navigators, the Phoenicians, exchanged wine for other goods somewhere around 1000 B.C. At first, their wine originated from central Asia-Minor, but later vineyards were planted on Crete, Cyprus and other Mediterranean islands. The earliest plantings of grapes in the region of Carthage in North Africa are also credited to this remarkable seafaring nation.

It is quite possible that the tribes of early Greece brought with them the knowledge and experience of wines in their migration from Asia. Homer, in the *Odyssey* and the *Iliad*, gives such a detailed description of vine-growing and winemaking that extensive wine production in his time cannot be disputed.

The first Greek settlers in southern Italy found so many native vines that they called their new home *Oinotria* — 'wineland'. Whether these vines were already cultivated by that time or were wild plants is not known, though Pliny, in his *Natural History*, writes that the Athenian, Eumolpus, was responsible

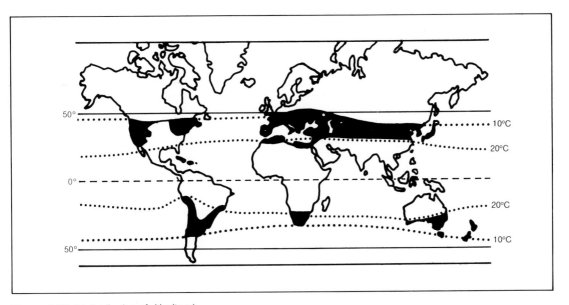

Figure 1 World distribution of viticulture*

*Note that virtually all quality wine is produced in the regions situated in the temperate-zones between the 10 °C, and 20 °C lines (these are the mean temperatures over the whole year)

for teaching Romans the cultivation of vines. As the vines in question, *Vitis vinifera*, are among the native plants of Italy, it is more than possible that some wines were made from these or the closely related *Vitis silvestris*.

Viticulture in Italy increased rapidly from south to north and, with the expansion of the Roman Empire, found its way towards the centre of Europe. The first vineyards planted in the south of France were those at Masillia (Marseilles) where viticulture started with the arrival of Greeks around 500 B.C. From here, the culture of vines progressed along the Rhône, but it was not until considerably later that vineyards were planted in other parts of France.

During the last century B.C., the Roman Emperor, Julius Caesar, conquered all of Gaul (France) and vines were planted in Bordeaux and Bourgogne. In Caesar's day the river Rhine was the border between the province of Gaul and the areas controlled by Teutonic tribes to the east— today's Germany — and, as had happened in other parts of their Empire, viticulture was encouraged by the Romans. Along the Moselle and Rhine, however, the first vines were most likely planted by the Gauls themselves.

It is interesting to note the contrast in vine-training methods used in the Moselle and Rhine valleys. The Moselle training of vines to a single post is rather similar to the style practised by early Greek growers, suggesting that vine-growing occurred before the Roman occupation: the trellised vineyards of the Rhine, however, show more obvious signs of Roman influence.

During the first centuries of this millennium viticulture in many parts of Europe was already well-established, largely due to the expansion of the Roman Empire, and wine grapes prospered in parts of France, Germany, Austria, Hungary and even England where, according to several reports, quality wines were made.

By the collapse of the Roman Empire, during the fifth century A.D., wine production had been mastered by the native populations in all the viticultural areas of Europe. From this point on the development of distinctive methods, in both vineyards and wine cellars, occurred in ways which were dependent on climate, soils, local conditions and the religious and socio-economic factors associated with each wine region.

More recently, with the colonisation of the New World, viticulture spread into all of today's wine-producing countries. In America, vines were first planted by Cortez in 1524, though native vines, *Vitis labrusca* and others, were al-ready present. In 1652, Jan van Riebeeck brought cuttings of grape vines to the newly-established province of the Cape of Good Hope, thus establishing the foundation of a wine industry in South Africa.

Viticulture in Australia and New Zealand owes much to wine enthusiast, James Busby —a botanist who planted vines in the early part of the nineteenth century.

The present world distribution of viticulture is shown in Figure 1.

An introduction to cool climates

Climate is undoubtedly one of the major factors determining both where grapes can be grown and the quality of the wine produced from those grapes. The lines on the map in Figure I indicate the boundaries between which most are produced. If the mean temperature is above 20°C, then winters are mild and leaf fall and vine dormancy do not occur or, if they do, occur only partially; temperate plants such as vines crop poorly under these conditions. If the mean temperature is below 10°C, summers are short and winters may be very severe. In short summers, vines have insufficient time to ripen their fruit and cold winters may kill or seriously damage the plant.

Quality

The temperatures quoted above are the extremes; within these limits vines may grow and crop, although not all produce grapes which are suitable for winemaking. In warm climates, the production of raisins, sultanas, currants, or bulk wines of mediocre quality are possible. As the temperature gets cooler, it becomes more difficult to produce dried fruit, but wines begin to have better quality and fortified wines are common; cooler still and fortified wines disappear from the cellars, but the quality of table wines improves dramatically. At the cooler limit, it is normal to find only white wines, but in selected areas in such districts, quality can be extremely high. Yields generally are higher in warm districts.

Cool districts, especially those near the cooler limit, pose a number of problems: frost may damage the crop in spring or autumn and, in cold summers, grapes may not ripen properly. Nevertheless, winemaking in districts near the cooler limit may be economic because, in the better years, quality is excellent and high prices may be achieved.

Temperature

It is not yet fully understood why cool climates generally produce the best quality table wines, but the evidence suggests that it is the lower temperatures in the autumn which are of special significance. In warm climates ripening of grapes occurs early, when the weather is still warm or even hot. These hot conditions cause rapid development of sugars, rapid loss of acids, and high pHs. The consequences of these developments will be discussed later, but at this stage it is sufficient to note that the juice is often unbalanced with respect to sugar, acid, and pH, and the grape appears to have had insufficient time to accumulate those many chemical compounds which add distinction to the wine. A cool autumn—often with considerable diurnal temperature variation — slows down development; better balances can be achieved, and more aroma and flavour constituents are accumulated.

Degree Days and LTI

One method of assessing the growth potential of a district is to measure the heat accumulated during the season. For this a base of 10°C (50°F) is taken because below this temperature very little vine growth occurs. The 'base' figure of 10°C is then subtracted from the average temperature for the month and the result is multiplied by the number of days in the month. If, for example, the average temperature* was 17.5 °C (63.5 °F), heat units would be (17.5 - 10) x 30 = 225 °C for a 30 day month, or (63.5 - 50) x 30 = 405 °F. The heat units (sometimes called degree days) for the growing season are added together to produce an assessment for the district. Heat units can be calculated on a daily basis which is more accurate and gives a figure somewhat higher than if done on a monthly basis. Data in this book are monthly calculations. It is possible to calculate average heat units for the previous twenty years or so to establish a fairly accurate picture for an area. Heat units can be useful for determining the suitability of various sites in a geographical area with similar climatic patterns.

*Average temperature equals mean daily-maximum for month plus mean daily-minimum for month divided by two .

They are less valuable when comparing different types of climate; for example, a maritime versus a continental climate.

Within one climatic zone it has been found that taking the mean temperature of the warmest month gives a result comparable to heat units. Such a method still has limitations when comparing areas with differing climatic types, however, its the simplicity has much to commend it.

Figure 2 shows the mean monthly temperatures for four grape-growing areas. Two areas are of special interest—Christchurch in the South Island of New Zealand and Geisenheim in Germany. Geisenheim has higher mid-summer temperatures and more degree days (1050°C-1890°F cf. 950°C-1710°F). This is primarily because it is more continental and has higher summer temperatures. Christchurch, on the other hand, is closer to the equator and has a longer growing season. If the range of varieties which can be grown in each district is compared, it is found that Christchurch has a greater ripening ability. In other words, the length of the growing season compensates for the cooler midsummer temperature.

The length of the growing is very dependent on latitude and, indeed, one study[3] suggests that latitude is a better indicator of climatic suitability than Degree Days. An even better guide is a system called the Latitude-Temperature Index (LTI).

LTI = mean temperature of the warmest month x (60- latitude). In the tables that follow, Degree Days and LTI are included. It will be noticed that, when comparing data between tables, LTI gives a much more accurate correlation between the grapes grown and climate.

What is a cool climate

For the purposes of the book, a cool climate is one which will have the capacity to produce table wines of distinction. In such areas, there will be variability in quality between seasons which will cause some to be labelled good vintages, some average and some poor. This is a typical characteristic of cool climates.

It is also possible to classify climates by the type of grapes that will ripen satisfactorily. In Table 1, four groups are identified; Group IA —cool; Group IB—cool to warm; Group IC —warm; Group II — warm to hot. Shown in the table are the LTI values which are

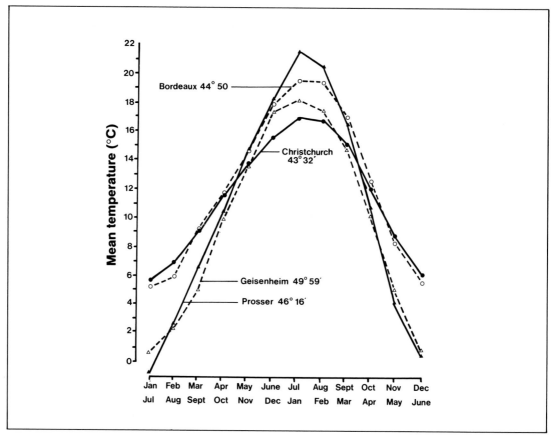

Figure 2 Mean monthly temperatures of four grape-growing districts: Bordeaux — France; Christchurch — New Zealand; Geisenheim—Germany; Prosser Washington USA

normally required for each group. Areas with higher LTI values ripen grapes in lower groups, but those with lower LTI values will usually find difficulty in ripening grapes of higher groups.

The cool-climate districts which will be described are those that ripen grapes in one or more of Groups IA to IC. They normally do not grow grapes from Group II. In terms of degree days, they are mostly below 1500°C or 2700°F and LTI below 380 (the Fahrenheit scale is not used for LTI).

Further aspects of climate will be discussed in Chapter 4, but one of these needs to be mentioned before the data in the next section on grape-growing areas is presented. Most quality grape-growing areas have rainfall below 750mm (29.5 inches) per annum. Wet areas, especially if the rainfall is concentrated in the autumn months, effectively lowers the length of the growing season. It will be seen from the tables in Chapters 1 and 2 that districts with 800mm or more rain per year tend to grow grapes in the group below the one predicted by LTI. High altitudes have similar effects.

Another limiting factor common to grape production is the degree of winter chilling. In very continental areas, such as central Europe, Asia, or North America, midwinter temperatures can be cold enough to kill or seriously damage the vines (this is different from frost in spring or autumn which damages green shoots and leaves). Winter injury is not common in most districts described in this book, but damage occasionally occurs in some European areas. In central Washington and Idaho, and many Eastern States and Canada, it is more common. Grapes in groups IC and II are generally more sensitive than those in groups IA and IB, and this can be another reason why, in some of these areas, grapes in lower, cooler groups are planted. Franco-American hybrids and native American grapes, are also more winter tolerant and are common in some of these districts.

Damage and death can occur if the winter temperature falls below -15 °C. The following

Table 1: Grape varieties grouped according to ripening ability in different climates

Group and LTI		Grapes grown
Group IA: LTI<190	1. very cool	Siegerrebe, Ortega, Optima, Madelaine x angevine 7672, Reichensteiner, Müller Thurgau, Seyval blanc, Huxelrebe, Bacchus.
	2. cool	Pinot gris, Pinot blanc, Pinot noir*, Pinot meunier*, Chasselas, Gewürztraminer, Sylvaner, Chardonnay*, Faberrebe, Kerner, Scheurebe, Auxerrois, Aligoté. Those grapes marked with an asterisk are especially suitable for producing Méthode Champenoise
Group IB: LTI 190-270	cool—warm	The key varieties are Riesling and Pinot noir—the latter can produce heavier, red wines, unlike the lighter wines made in Group IA, Chardonnay in such districts is more full-bodied also.
Group IC: LTI 270-380	warm	The key varieties are Cabernet Sauvignon and related Cabernet Franc, Merlot, Malbec, plus Sauvignon blanc and Sémillon. These are sometimes grown in the cooler regions of Group IB, but seldom reach the same quality.
Group II: LTI 380 and above	warm—hot	Carignane, Grenache, Shiraz, Thompson's Seedless (Sultana), Cinsaut, Zinfandel once again, some of these will ripen in Group IC districts, but their cultivation is mostly restricted to warm to hot climates

guide will indicate the likelihood of this occurring. If, after examination of meteorological records, the mean temperature of the coldest month is seen to be below -1°C; or if the lowest temperature for a 20-year period reaches -20 °C more than once, then the climate will probably be marginal for grapes.[1]

One further point needs to be made and this is in relation to an older classification of regions according to Degree Days,[7]. Five regions, are specified: I, cool (below 1390 °C or 2500 °F), to V, hot (above 2220 °C or 4000 °F days). For such a broad classification, degree days are satisfactory, but cool-climate grape-growing areas virtually all fall into Region I and, for this group, Degree Days, as noted earlier, are misleading. Groups IA, IB, and IC, are in fact, cool-climate sub-divisions of Region I climates, using a more accurate method — the LTI system. Regions in this book are therefore classified as regions IA, IB, IC, II, III, IV, V.

References and further reading

[1] Becker, N., 1985. Site selection for viticulture in cooler climates using local climatic information. Proc. 1st Int. Symp. Cool Climate Vitic. and Enol. Oregon State University, Corvallis.

[2] Coombe, B.G., Dry, P.R. 1988. *Viticulture Volume I, Resources in Australia.* Winetitles, Adelaide.

[3] Jackson, D.I., Cherry, N.J. 1988. Prediction of a district's grape ripening capacity using a latitude temperature index (LTI). *Amer. Jour. Enology & Viticilture,* 39: 19-28.

[4] Gladstones, J. 1992. *Viticulture and Environment.*
Winetitles, Adelaide.

[5] Pongràcz, D.P. 1978. *Practical Viticulture.* David Philip, Cape Town.

[6] Weaver, R.J. 1976. *Grape Growing.* Wiley, New York.

[7] Winkler, A.J., Cook, J.A., Kliewer, W.M., Lider, L.A. 1974, *General Viticulture.* Univ. Calif. Press, Berkeley.

CHAPTER 1

FACTORS CONTRIBUTING TO QUALITY IN SOME COOLER EUROPEAN WINE DISTRICTS

The selection of areas in this part of the book is not intended to be all inclusive and some well-known wine-producing countries have been omitted. The prime intention is to show, by example, the climates, soils and varieties that have been successfully used to produce fine wines in selected cool-climate grape-growing districts.

France

In terms of quantity and quality, France is a leader in world wine production. Generally, it is the warmer districts of the south which produce the quantity and the cooler areas of middle France and the north which contribute most to quality.

French wine production is controlled by the laws of *Appellation d'Origin* and *Appellation Contrôlée*. The effect of these strict, though somewhat out-dated, laws on the volume and quality of local wines is sometimes not fully appreci-

ated outside France. The *Appellation d'Origin* limits the production of certain types of wines to recognised wine districts — Champagne, Burgundy, Chablis and so on. In most cases it also specifies what varieties of grapes are to be used. These are the grapes that can be relied upon to produce quality wines and contribute to the authenticity and overall quality of French wines. The *Appellation Contrôlée* label implies even stricter controls in both vineyard and wine cellar. For example, only some 60 per cent of Burgundy

St Emilion

Table 1.1 The major wines, varieties, soils and climates in quality wine districts of France

District	Wines	Varieties
Champagne	Sparkling and still white	Pinot noir, Pinot meunier and Chardonnay
Alsace	White and rosé	Chasselas, Sylvaner, Pinot gris, Pinot blanc, Gewürztraminer, Muscat blanc, Riesling, Muscat Ottonel, Auxerrois and Pinot noir
Chablis	White	Chardonnay and Aligoté
Loire	White, red, rose, and some sparkling	Chenin blanc, Melon, Sauvignon blanc, Chasselas and Cabernet Franc
Côte d'Or	Red and white	Pinot noir, Chardonnay, Aligoté and Pinot blanc
Chalonnaise	White and red	Pinot noir, Gamay noir, Chardonnay. Aligoté and Pinot blanc
Mâconnais	Red and white, some sparkling	Pinot noir, Gamay noir, Chardonnay, Aligoté and Pinot blanc
Beaujolais	Red	Gamay noir
Côte Rôtie	Red	Syrah and Viognier
Hermitage	Red and white	Syrah, Marsanne and Rousanne
Châteauneuf-du-Pape	Red and white	Grenache, Syrah, Cinsaut, Clairette, Mourvèdre, Rousanne, Grenache blanc and others.
Tavel / Lirac	Rosé and red	Grenache, Cinsaut, Syrah and Mourvèdre
Médoc	Red	Cabernet Sauvignon, Merlot, Malbec, Cabernet Franc, Petit Verdot, Saint-Macaire and Carmènere
Graves	Red and white	For red wines as in Médoc, for white: Sémillon, Sauvignon blanc and Muscadelle
St Emilion and Pomerol	Red	Cabernet Franc, Merlot, Cabernet Sauvignon, Malbec (Cot), Saint-Macaire.
Sauternes	White	Sémillon, Sauvignon blanc and Muscadelle

wines are entitled to this label in an average vintage. Its main restrictions include the yield of wine per hectare, the minimum alcohol level of wines, the grape varieties and their composition in the vineyard, the use of fertilisers, the time of harvest, and other factors relevant to wine quality. Bottled wines are then checked both chemically and by taste to maintain the reputation of wines produced under this label.

The abbreviation V.D.Q.S. (*Vins Délimités de Qualité Supérieure*) represents yet another regulation, but this is less restrictive than the *Appellation Contrôlée*.

Even the finest of French wines can be declassified in lesser vintages when failing to meet the strict quality standards set by their appellation. For example, wine from Château Margaux, ie: the *Premier Grand Cru Classé* vineyard in the commune of Margaux, of the Bordeaux region Haut Médoc, can be declas-

Soil Type and Position	Annual Rainfall	Heat Units in Growing Season* (base10°C, 50 °F)	LTI
Chalky soil on hard rocky subsoils Flats and low slopes	700 mm (28 in)	1050 °C (1890 °F)	191
Gravelly loams of volcanic or granite nature, often rich in limestone. Low slopes	650 mm (26 in)	1230 °C (2210 °F)	219
Thin layer of loam over grey-white subsoils, rich in limestone. Flats and gentle slopes	700 mm (28 in)	950°C(1710°F)	228
Volcanic, rocky soils on slopes in north; lighter, sandy loams, often rich in limestone in the central valleys. Clay, alluvial gravelly loams in the southern parts Mostly on slopes	620 700 mm (24-28 in)	950-1100 °C (1710-1980 °F)	236
Clay loams rich in ironstone, in parts marl soils rich in chalk Alluvial, gravelly-silt or volcanic rock subsoils Flats and gentle slopes	630 mm (25 in)	1180 °C (2120 °F)	263
Gravelly loams, alluvial soils over rocky subsoils Flats	680 mm (27 in)	1120°C(2020°F)	266
Clay loams with outcrops of volcanic rocks, in parts rich in limestone Low slopes	680 mm (27 in)	1150 °C (2070 °F)	283
Clay loams over volcanic rock and granite subsoils Flats and slopes	700 mm (28 in)	1150 °C (2070 °F)	290
Thin layer of dark loam over rocky hard subsoils Very steep slopes and terraces	530 mm (21 in)	1400 °C (2520 °F)	356
Stony loams over volcanic, rocky subsoils. Steep slopes, terraces	550 mm (22 in)	1450 °C (2610 °F)	378
Poor stony soils over hard, granite subsoils. Flats and gentle slopes	530 mm (21 in)	1500 °C (2700°F)	385
Stony loams over granite and rocky subsoils. Flats	530 mm (21 in)	1450 °C (2610 °F)	370
Alluvial, clay loams often rich in gravel and quartz sand. Flats and gentle slopes	750 mm (30 in)	1350-1400 °C (2430-2520 °F)	300
Gravelly soils in the northern parts, more silt and clay loams in the south. Flats and gentle slopes	720 mm (28 in)	1250-1300 °C (2250-2340 °F)	303
Gravelly, stony soils on slopes in central part, lighter and better drained, alluvial sandy soils elsewhere. Clay and silt loam subsoils are common in flats. Gravelly loams over compact, clay loam subsoils in Pomerol Flats and mild slopes	700 mm (28 in)	1250-1300 °C (2250-2340 °F)	303
Gravelly loams rich in limestone in the north, marl soils over stony subsoils in the south. Mostly flats	650 mm (26 in)	1350-1400 °C (2430-2520 °F)	315

See Historical Introduction for the definition of heat units and LTI

sified into 'Margaux AC', or further to the lower ranks of wines labelled: 'Haut-Médoc' or 'Bordeaux Supérieure'. The less precise geographical designation on a label of French wine can imply such a declassification.

Total average wine production in France amounts to an impressive 8000 million litres (1700 million gallons) produced annually from over 1.2 million hectares (3 million acres) of land. However, it should be remembered that not all French wine is of high quality and about half is at best only average, such wine being sold as *Vin Ordinaire* or *Vin du Table*.

The recognised areas produce fine wines every year. In the most favourable vintages their quality and distinction can be matched only by a few from anywhere in the world. The best-known classical districts of France are: Alsace, Champagne, the red and white wine-producing areas of Burgundy, Bordeaux and parts of the Côtes-du-Rhône and Loire.

The lesser table wines are produced in the large districts of the Midi, the Gard, Hérault and Aude; in parts of the Rhône valley, the Pyrénées-orientales and elsewhere in southern France.

The quality of wine produced in France, as elsewhere, is dependent on climate, soil, varietal composition, seasonal variations in yield and ripeness, the winemaker, his/her equipment and ability, and finally, the strict wine laws that have already been mentioned.

While the standard of viticulture in France is generally very good, assisted by the numerous I.N.R.A. research stations found in all regions,

the winemaking techniques are often of a surprisingly low standard. There are well-equipped cellars in all the French wine districts, but many are badly in need of modernisation in order to make French wines more competitive on the world markets. The increased demand for the better wines of Chile, California, South Africa, Australia, and New Zealand indicates that these wines have not only achieved fine quality and consistency but, more importantly, they are being produced more cheaply than in France.

Table 1.1 and Figure 1.1 summarise classical districts, important wine types and varieties, together with the geography, soils and climates in which they grow.

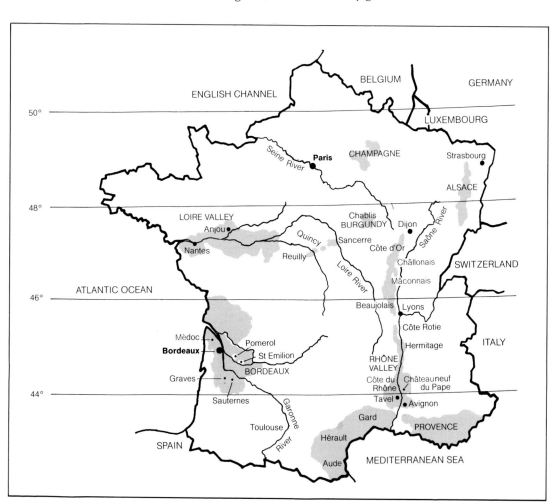

Figure 1.1 The major wine regions of France

Germany

The less reliable and cooler climates of Germany forced early winemakers to study every possibility of improving their vineyards to gain the maximum maturity in grapes for quality wine production. This centuries-old effort has influenced wine production in all German wine districts to such an extent that wines are often said to be produced in the vineyards, rather than in the cellar. Careful selection of the vineyard site according to soil type and microclimate,

together with understanding, dedication and hard work in the vineyard have, with an expert knowledge in winemaking, enabled fine wines of distinction and style to be made.

Germany is mainly a white-wine producing country, although some red wine (16%) of fair quality is made, especially in the southern areas. The area in production is over 96,000 hectares. The most important grape varieties grown include, in order of decreasing importance: Müller-Thurgau, (Riesling Sylvaner),

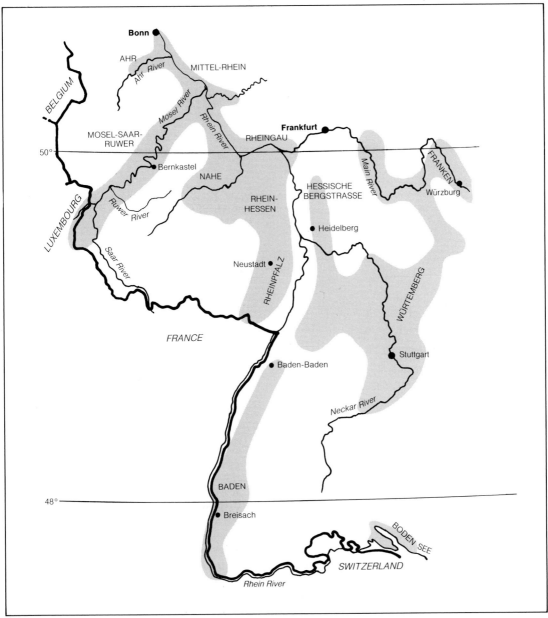

Figure 1.2 The major wine regions of Germany

Table 1.2 The major wines, varieties, soils and climates of Germany*

District	Wines	Varieties (in order of importance)
Baden	White (76%) and red	Müller-Thurgau, Pinot noir, Pinot gris, Chasselas, Portugieser, Riesling, Sylvaner, Pinot blanc and Gewürztramin.
Rheinpfalz	White (92%) and red	Müller-Thurgau, Riesling, Sylvaner, Kerner, Morio-Muscat, Portugieser, Scheurebe, Pinot gris, Bacchus Gewürztraminer, Faber, Pinot blanc
Rheinhessen	White (96%) and red	Müller-Thurgau, Sylvaner, Scheurebe, Bacchus, Faber, Kerner, Riesling, Morio-Muscat, Portugieser, Pinot gris and Pinot noir.
Hessische-Bergstrasse	White (99%) and red	Riesling, Müller-Thurgau, Sylvaner and Pinot gris
Würtemberg	White (53%) and red	Riesling, Trollinger, Müllerrebe, Müller-Thurgau, Kerner, Sylvaner, Limberger, Portugieser, Pinot noir and Pinot gris.
Franken	White (98%) and red	Müller-Thurgau, Sylvaner, Bacchus, Kerner, Scheurebe, Riesling, Gewürztraminer and Pinot noir.
Rheingau	White (96%) and red	Riesling (75%), Müller-Thurgau, Pinot noir and Kerner
Nahe	White (98%) and red	Müller Thurgau, Riesling, Sylvaner, Kerner, Scheurebe, Bacchus, Faber, Pinot gris, Pinot blanc and Morio-Muscat
Mosel-Saar-Ruwer	White (100%)	Riesling (58%), Müller-Thurgau, Elbling, Kerner, Bacchus and Scheurebe.
Mittel-Rhein	White (99%) and red	Riesling, Müller-Thurgau and Kerner
Ahr	Red (55%) and white	Pinot noir, Portugieser, Riesling and Müller-Thurgau

*The extensive use of slopes in German vineyards creates meso-climates to increase heat accumulation The more heat-demanding Riesling is grown on the warmest sites in a district where LTI will be above the figure given.

Kerner, Scheurebe, Pinot noir, ie: Blauer Spätburgunder, Pinot gris, ie: Rülander, Bacchus, Portugieser , Morio-Muscat , Faber, Trollinger and Chasselas, known also as Gutedel, and Elbling.

As is typical of cooler climates, yield and quality vary considerably. For example, in the last ten years the yield has been as low as 463,500 litres (101 900 Imp. gallons, 122 440 US gallons) in 1980, and as high as 1.5 million litres in 1990.

Generally, the finest vineyard sites are found on slopes facing south, south-east or south-west, with individual vines trained to a post or trellis. The vineyards often follow the paths of rivers, and soils—often containing gravel or stones — are invariably well-drained and of medium to low fertility. All these aspects not only give German vineyards their picturesque outlook, but also contribute to the quality of grapes produced.

Vines tend to be trained on moderately high trellises or posts which allow better exposure to sunlight and a greater movement of air among the ripening grapes. This is also important in spring when late frosts can

Soil Type and Position	Annual rainfall	Heat Units in Growing Season	LTI
Clay loams, in parts gravelly soils over alluvial subsoils. Flats	670mm (26 in)	1050 °C (1890 °F)	222
Light, sandy loams on rocky subsoils, gravelly and rich in basalt (Pechstein) in the best parts Clay loams of greater fertility can be found in the south. Flats and gentle slopes	700-750mm (28 -30 in)	1200-1250 °C (2160-2250 °F)	201
Varies considerably, from heavy, fertile, clay loams—sometimes rich in limestone—in central parts, to well-drained sandstone or chalky soils over subsoils of marl along the Rhine river	700mm (28 in)	1050-1150 °C (1890-2070 °F)	188
Alluvial loams, with gravel, in parts rich in limestone. Mainly slopes	650mm (26 in)	1150 °C (2070 °F)	192
Clay loams, stony soils in parts over silt subsoils. Mainly slopes	650mm (26 in)	1150-1200 °C (2070-2160 °F)	194
Marl loams over alluvial subsoils in the east, the central part has chalky soils, rich in limestone. Sandy loams are found in the western part. Mainly slopes	650mm (26 in)	1050-1150 °C (1890-2070 °F)	186
Varies from brown-coloured calcareous loams to heavier, silt loams over stony or loess subsoils. Alluvial limestone soils can also be found. Mainly slopes	650-700mm (26-28 in)	1050-1200 °C (1890-2160 °F)	188
Clay and silt loams over loess subsoils in the north and the north-east, more stony or even slaty soils over volcanic rock subsoils in the west. In the best areas soils rich in ironstone and copper are common. Mainly slopes	600mm (24 in)	1100-1200 °C (1980-2160 °F)	192
Chalky soils over rocky subsoils in the north; the best parts of central Mosel have slate soils on rocky subsoils. In the Saar, the soils vary from slate in the north to limestone in the south. In Ruwer, slate soils prevail. Slopes	600-650mm (24-26 i n)	950-1150 °C (1710-2070 °F)	184
Slate soils over volcanic rock subsoils in the better parts, elsewhere loess with pockets of chalky soil can be found. Slopes	650mm (26 in)	1050-1100 °C (1890- 1980 °F)	180
Volcanic, rocky soils on loess or stony subsoils. Slopes	620mm (24 in)	900-1100 °C (1620-1980 °F)	175

damage the vine shoots. When a vineyard is planted on a slope, high training on the trellises permits the cold air to drain to the bottom of the valley and keep the vineyard relatively frost-free. Later, in the autumn, air circulation among the vines has a favourable effect on the development of flavour, aroma and sugars in the slowly-maturing grapes. Well-drained soils, especially if stony and clear of vegetation, retain the heat of the day well into the night, thus increasing the ripeness in grapes, though the contrast of day and night temperatures is important as well. Furthermore, vines planted on well-drained slopes are not only better exposed to the sun, but when the soil dries out in the warm, autumn weather which is sometimes experienced, grapes may shrivel on the vines and the sugar, flavour and aroma is further concentrated. Thus the temperature and drainage factors — which result in major German wine areas having grapes with a slower and more complex maturation pattern— are largely responsible for the distinctive character of the better wines.

German winemakers appear to have adopted modern winemaking technology with greater enthusiasm than the French, and well-equipped cellars are found in most regions.

The increase in co-operative wine production and the rapid development of new technology is evident in all aspects of German winemaking and has resulted in competitive pricing of German wines around the world.

The large teaching and research contribution made by universities, wine colleges and research institutes has also been important in assisting the development of wine quality and wine making expertise.

A further beneficial contribution has been made by the revised German Wine Law, which was enacted for the purpose of simplifying the labelling and nomenclature of local wines and, more importantly, for clarifying the regulations which govern wine production.

Where the more traditionally-orientated French regulations classify wine areas and vineyards alone, the German law permits any wines of any district to reach different levels of quality, providing they meet the requirements set for both vineyard and cellar. Under the law of 1971 the development of wine areas and grape types is further encouraged.

Figure 1.2 shows the distribution of the important grape-growing areas in Germany.

The categories used are mentioned in

Vineyards in the Mosel. Vines are grown on poles 1 m (3 ft) apart on these valuable slopes The slate soils, absorbing heat during the day and releasing it at night, good drainage and the southerly aspect, create a most desirable meso-climate that allows the Riesling grape to grow to perfection in many years

Chapter 12. Variability in the climate is reflected in the qualities of the various types made; for example, in 1983, 82 per cent of wine was labelled 'Qualitätswein mit Prädikat' (QmP) and average sugar levels were 84° Oechsle or 20° Brix (see Table A.1.2 in Appendix 1), in 1984 wines averaged 60° Oechsle and only 12 per cent reached the QmP level of quality.

The world-wide popularity of low-alcohol, slightly-sweet, white table wines is reflected in many German wines having sweet reserve, i.e. grape juice, added prior to bottling (see Chapter 12). This trend, over the past thirty years, has changed to a pattern of greater wine consumption with food, with the result that it has reduced the amount of residual sweetness required by the market. Such drier types are usually imported into Germany from France, Italy and Spain. But it is of interest to note that the number of sugar-free, dry or only slightly sweet wines produced in Germany has increased markedly in recent years and labels describing the wine as halb-trocken or trocken (i.e dry) are seen with increasing frequency.

Table 1.2 indicates that, in Germany's wine districts as in France, rainfall tends to be rather low and that soils — while having perhaps more loam/clay particles—are well drained, often have high stone content, and sometimes contain a moderate limestone content. Total heat units are lower than in most French districts and sufficient ripening can be a problem in cold years.

Switzerland

The history of Swiss viticulture is closely linked to that of its chief neighbours — France, Germany and Italy.

In the French speaking part of the country, 'La Suisse Romande', vine growing dates back to the Roman times. The priory of Satigny near Geneva, built in AD 50, has been closely associated with viticulture since the early tenth century.

In most regions the traditional Roman way of vine training to tall stakes is practised to this day, and many of the ancient vine cultivars grown here were known to the Romans. The popular 'Amigne' was known two thousand years ago as *Vitis aminea*, and 'Muscat' was called by the Romans *Vitis apiana*.

In more recent times, the Fendant or Chasselas was introduced from France at the time of the reign of Louis XV. Pinot noir vines, said to have been planted in 1848, produced the famous red 'Petit Dôle' wine of Valais for the first time in 1851. The introduction of the Gamay grape from Beaujolais, and Marsanne from Hermitage—plus the 'champagne-like' century-old practice of secondary bottle fermentation of white wines near Neuchâtel —all clearly point to the close association of early Swiss grapegrowers with their counter-parts in France.

Similar developments can be seen in the Ticino district near the Italian border, where the Italian traditions and wine grapes dominate The exception here is the Merlot grape, which originated in Bordeaux.

The influence of German viticulture on wine production in the northern part of Switzerland near Zürich and the Schlaffhausen districts is also of interest. The diversity of Swiss viticulture is seen from the fact that, throughout the German-speaking part of Switzerland, Pinot noir, known as Klevner or Blauburgunder, is grown and made into a traditionally French style of red wine. Traditional German grapes—Sylvaner and Müller-Thurgau and ever-present Chasselas—are predominantly used for making white wines.

The Swiss Wine Law focuses much attention on the varietal composition of grapes grown in each district, yet, rather surprisingly, must sugaring (chaptalisation) is not controlled; yields are not limited as in France, and wines are not graded according to the natural ripeness level reached, as is common in Germany or Austria. One logical reason for this must be the high cost of producing wine in a country where only high yield and the mechanisation

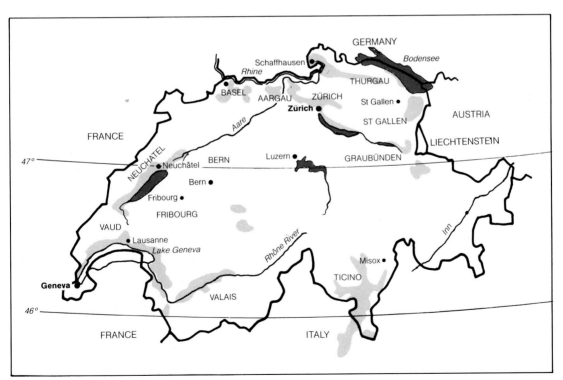

Figure 1.3 The major wine regions of Switzerland

Table 1.3 The major wines, varieties, soils, and climates in Switzerland

Region	Districts and wine types	Varieties
Westschweiz (Western Region) 72% white wines	Valais, 64% white wines Waadt, 83% white wines Geneva, 64% white wlnes Neuenberg, 73% white wines Bern, 81% white wines Freibourg, 88% white wines	Chasselas, Sylvaner Pinot noir, Gamay, Syrah, Pinot gris, Amigne. Arvine, Réze, Humagne, Muscat blanc, Heiden (Päien)
Ostschweiz (Eastern region) 73.5% red wine	Zürich, Schaffhausen with a smaller amount produced in Thurgau . St Gallen, Graubünden, Aargau, and Liechtenstein	Pinot noir, Müller-Thurgau (often called Riesling Sylvaner in Switzerland), Chasselas, Rauschling, Pinot gris, Gewürztraminer
Südschweiz (southern region) 99% red wines; 20% of vineyards are in hybrid grapes	Ticino and Misox districts	Wine grapes: Merlot, Bondola, Grape juice: Isabella, Clinton, Noah

of vineyard work can achieve the returns that are expected in the vineyards of France, Italy, Spain, and Germany — the four major suppliers of wine to the Swiss population.

The Swiss wine regions are closer to the equator than the German viticultural districts. Nevertheless, the higher altitude and the higher rainfalls recorded in many Swiss vineyards gives them cooler, and often more marginal climates. For this reason, the choice of soils and grape types is crucial and of the utmost importance. The soils are well drained, and are often of a calcareous, schist or stony nature, so that they are easily warmed by the sun. In addition, slopes facing south, south-west and south-east are preferred, in order to improve the accumulation of heat and to increase the light received by the vines whilst, at the same time, minimising the danger of spring frosts.

The use of steep slopes, by terracing, is well advanced in all Swiss regions, and terraces for mechanised vineyard operations are common.

The less than 14 000 hectares of vineyards produce some 107 million litres of wine of which 61 per cent is white wine. A small amount of 'Franco-American' hybrids are grown and these are chiefly used for the production of excellent grape juices, or as table grapes.

The largest wine producing area is the western, French-speaking region, (Westschweiz). Due to cool temperatures the local white wines contain only moderate amounts of alcohol and even the better vintages are chaptalised (sugared). The widespread use of de-acidification by natural or induced malo-

lactic fermentation, or chemicals in white and red wines, is common. Most of the wines made are bottled early and can be quite gassy. White wine is often bottled under carbon-dioxide pressure, as many appear a little too soft and low in acidity after the de-acidification used earlier in the winemaking. Most of the Swiss

Terraces in Swiss vineyards. Photography by courtesy *Wädensville Wine Research and Teaching Institute.*

Soil types and position	Annual rainfall	Altitude above sea-level	Degree Days	LTI
Low slopes. 5-15% inclination, south to south-east exposures. Steepest vineyards, up to 30% inclination, face south or south-west. Soils mostly calcareous gravels, chalky in parts and often with high schist/stone content. Lower-placed vineyards have soft, clay soils deposited over glacial sedimentation	600-1150mm (23-45 in)	450-650m (1440-2130ft)	850-1100 °C (1530-1980 °F)	230-250
South to south-east facing slopes reaching up to 25-30% inclination. Soft, marl soils often rich in silica on slopes. Alluvial, calcareous loams over grey marl subsoils in lower positions	850-1300mm (33-51 in)	450-500m (1440-1640ft)	850-1 000°C (1130-1800 °F)	200-230
Low slopes and flats, inclination of slopes reaches 5-15% facing south or south east. Soils are gravelly loams mixed in parts with soft, calcereous marl	1700-1800mm (67-71 in)	225-400m (740-1310ft)	1150-1250°C (2070-2250 °F)	250-260

white wines appear best while young, retaining at that time their freshness and delicate bouquet.

A considerable amount of red wine from Gamay and Pinot noir grapes is also produced in the Valais district. This is made and marketed in a manner similar to French Beaujolais, to which it favourably compares. Quality red wine production is also a characteristic of the eastern Swiss districts (Ostschweiz). The Pinot noir dominates in this area and traditional German grapes account for the rest.

The smallest of the Swiss regions, Südschweitz, centres on the Ticino district where soft, tannic red wines are made from Merlot and Bondola grapes. Less than one per cent of the wine made here is white, for which Chasselas and Semillon are grown. Due to higher heat accumulation, the red wines of Ticino are more robust and fuller flavoured than other Swiss red wines. In view of this, and given the popularity of the French red wines on the Swiss market, it is rather surprising to find that the wines of Ticino are less well-known and that more of them are not exported out of the region than the wines of Westschweiz.

The maximum use of steep terrain, cool-climate and modern winemaking techniques observed in Switzerland today, would not have been possible without the development of advanced research and teaching facilities that are available to Swiss vinegrowers. The viticultural and wine research is co-ordinated from the Wädensville Institute, and wine schools are established in all the major regions.

Austria

The grape vine was cultivated in Austria prior to and during Roman times, however, little is known of the wine quality produced in that period. After the departure of the Romans, the original vineyard area declined and no significant amount of wine was made until the tenth century when, in AD 955, Otto I encouraged vine planting and large wine cellars were built by the abbeys near Salzburg. By the late sixteenth century, the viticultural regions of Austria amounted to an impressive 150 000 hectares.

The first Wine Law was established in 1673. All existing vineyards were then classified according to records of production, which had been kept for some time prior to this date. Wines were divided into three major groups; good, fair, and ordinary, depending on their reputation for quality and their price. During the reign of the Empress Maria-Theresa in the eighteenth century, more detailed regulations for the wine trade were introduced, and by 1780 the Austrian Wine Law controlled the storage and shipping of wine as well as its production. The whole of the eighteenth century marked an increase in both the production and the quality of the wines which were made—that century also saw the introduction of two most important red grapes — Blauer Portugieser (1772) and Pinot noir (1800).

The great decline of the vine-growing areas in Austria at the end of the nineteenth century can be chiefly attributed to the appearance of phylloxera, and the introduction of 'Franco-American' hybrids which followed. By 1912, some 90 per cent of all Austrian vineyards were affected: the vineyards of Eastern Burgenland planted in sandy soils being the only exception.

Due to its continental, cool climate, the Austrian vineyards produce mainly white wines (84 per cent), though quality red wine is also made. An average vintage produces some 3 million litres (793 000 US gallons) of wine from the 60 000 hectares (148 000 acres) of vines grown. Wine production from 'Franco-American' hybrid grapes — known here as

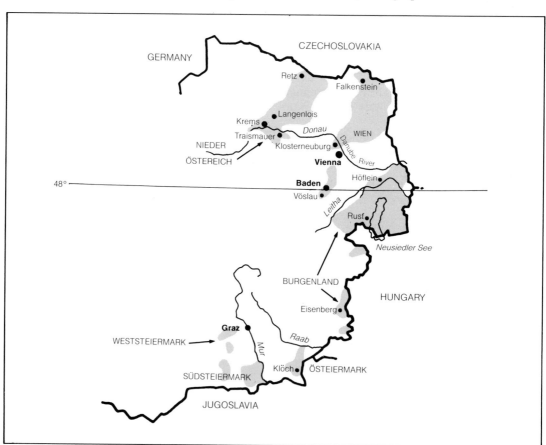

Figure 1.4 The major wine regions of Austria

Vineyard with Pinot noir This system with small terraces (1.5 m) makes viticulture possible on steep slopes. Photography by courtesy of *The Klosterneuburg Wine Research and Teaching Institute*.

(0.5 per cent) and other red grapes, amounting to 1.2 per cent of the total.

As in other cool-climate, central-European wine regions, the quality of Austrian wines depends on vintage and on district variations. The better wines are made from grapes grown on slopes exposed to the south and south-west, and in vineyards planted in soils that are well drained and contain a high proportion of schist, loess, or stones — often with a high chalk content. As with Swiss vineyards, high rainfall in some districts means that, in Austria, they ripen lower (cooler) group grapes than expected.

The modernisation of Austrian wine production, including the valuable research into viticulture and winemaking at the Klosterneuburg Teaching Institute near Vienna, has also played a part in the revival of quality wine production during the last three decades. The introduction of a detailed new Wine Law in 1969 is also considered important.

The best Austrian wines are made from quality varieties and are, under the new Wine Law, further classified into sub-groups according to ripeness. The ripeness is determined by the must weight measured in degrees on a 'Klosterneuburg Must Hydrometer' (°KLM), the equivalent in °Brix is given in Table 1.4.

Table 1.4 Wine classes in Austria

Wine Class	Minimum required ripeness	
	°KLM	°Brix
Quality wine (Qualitätswein)	15°	17.6
Certified merit (Prädikatswein)		
Spätlese	19°	22.3
Auslese	21°	24.6
Eiswein	22°	25.8
Beerenauslese	25°	29.3
Trockenbeerenauslese	30°	35.2

All quality Austrian wines bear the name of the grape, district, and, where applicable, the vineyard name. The lesser, bulk wines are never exported and usually are sold in the cafes or wine bars of larger towns and villages.

'direct producers'—has been discouraged by the Wine Law since 1963, although as much as 1000 hectares (2400 acres) had been planted as recently as 1961.

The quality, white wine grapes of Austria include: Grüner Veltliner (32.1 per cent), Müller-Thurgau (10 per cent), Welschriesling (7.7 per cent), Weissburgunder (Pinot blanc) (3.4 per cent), Neuburger (3.3 per cent), Riesling (2.1 per cent), Veltliner-Frührot (2.1 per cent), Muscat Ottonel (2.0 per cent), Traminer (1.6 per cent), Veltliner-Braunrot (1.1 per cent), Bouviér (1.0 per cent). Other white wine grapes, plus table grapes, constitute the remaining 17.3 per cent.

The red grapes include: Blauer Portugieser (5. 1 per cent), Blaufränkisch (4.3 per cent), Zweigeltrebe (4.1 per cent), Pinot St Laurent (1.1 per cent), Blauer Burgunder (Pinot noir)

Table 1.5 The major varieties, wines, soils and climates in Austria

Region	District	Wines	Varieties	Soil type and position	Annual rainfall	Heat units	LTI
Niederösterreich	Krems	96% White	Grüner Veltliner, Müller-Thurgau, Riesling, Neuburger, Zweigeltblau, Pinot St. Laurent.	Mainly loess and sandy loams on terraced slopes.	528 mm (21 in)	1150-1200°C (2070-2160 °F)	216
	Langenlois	94% White	Grüner Veltliner, Müller-Thurgau, Riesling, Veltliner-Frührot, Neuburger, Roter Veltliner, Zweigeltrebe and Blauer Portugieser.	Loess or brown loams over rocky sub-soils. Terraced slopes or on lower positions on valley floors.	534 mm (21 in)	1100-1150 °C (1980-2070°F)	212
	Retz	85% White	Grüner-Veltliner, Müller-Thurgau, Welschriesling, Weissburgunder, Veltliner-Frührot, Blauer Portugieser, Zweigeltrebe.	Alluvial loams of moderate to high fertility, granite sub-soils in low positions, South facing low slopes are also planted.	500-600 mm (20-24 in)	1150-1250°C (2070-2250°F)	215
	Falkenstein	94% White	Grüner Veltliner, Welschriesling, Müller-Thurgau, Weissburgunder, Blauer Portugieser and Pinot St. Laurent.	Low, terraced slopes and flats, Alluvial loams or loess of moderate to high fertility.	600mm (24 in)	1050-1200°C (1890-2160°F)	200
	Traismauer Carnuntum	90% White	Grüner Veltliner, Müller-Thurgau, Veltliner-Frührot, Weissburgunder.	Terraced, low slopes and flats. Coarse gravels, loess, in parts rich in chalk.	600-700mm (24-28 in)	1050-1150°C (1890-2070 °F)	205
	Baden	84% White	Neuburger, Zierfandler, Rotgipfler Weissburgunder, Grüner Veltliner, Veltliner-Frührot, Blauer Portugieser.	Terraced, low slopes and flats. Loess, gravels, in part rich in chalk.	700mm (28 in)	1000-1150°C (1800-2070 °F)	200
	Vöslau	66% Red	Blauer Portugieser, Pinot St. Laurent; smaller amounts of Blaufränkischer, Zweigeltrebe, Neuburger, Müller-Thurgau and Weissburgunder.	Terraced low slopes and flats, sandy loams, gravel soils (Steinfeld), in parts rich in silica or chalk	650-700 mm (26-28 in)	1150-1250°C (2070-2250°F)	202

Region	District	Wines	Varieties	Soil type and position	Annual rainfall	Heat Units	LTI
Wien (Vienna).		91 % White	Grüner Veltliner, Riesling, Weissburgunder, Müller-Thurgau, Blauburgunder, Pinot St Laurent, Zweigeltrebe.	Mostly on terraced slopes, with loess soils, in parts rich in chalk and limestone.	640 mm (25 in)	1100-1150°C (1980-2070°F)	234
Burgenland	Rust/ Neusiedlersee	81 % White	Grüner Veltliner, Welschriesling, Müller-Thurgau, Muscat-Ottonel, Neuburger, Weissburgunder, Gewürztraminer, Riesling, Bouvier, Blaufränkisch, Pinot St Laurent, Zweigeltrebe.	West facing, low slopes and flat positions. Sandy loams.	630-660 mm (25-26 in)	1200-1300°C (2160-2340°F)	220
	Eisenberg	70% White	Welschriesling, Riesling, Blauburgunder, Blaufränkisch.	South facing, terraced slopes, stony soils in Pinkatal valley and sandy loams on flats.	670 mm (26 in)	1050-1100°C (1890-1980°F)	217
Steiermark.	Klöch - Oststeiermark	96% White	Welschriesling, Müller-Thurgau, Gewürztraminer, Weissburgunder, Zweigeltrebe.	Volcanic soils in parts rich in basalt. South facing slopes are terraced.	880 mm (35 in)	1150-1200°C (2070-2160°F)	247
	Südsteiermark	96% White	Welschriesling, Müller-Thurgau, Gewürztraminer, Weissburgunder, Sauvignon blanc (Muscat-Silvaner), Riesling, Blauer Wildbacher, Scheurebe.	Steep slopes high above sea level, terraced, with schist soils of high stone content. In parts soils are rich in chalk.	950 mm (37 in)	1200°C (2160°F)	258
	Weststeiermark	Red & White 50% each	Blauer Wildbacher, Weissburgunder, Müller-Thurgau, Scheurebe	Steep, terraced slopes with calcereous soils of sedimented chalk and limestone. In parts rich in schist and stone .	1050 mm (41in)	1050-1150°C (1890-2070°F)	220

England and Wales

Britain is not generally regarded as a wine-producing region and, in fact, until recently vines were only seen in greenhouses or conservatories. However, large numbers of growers have become established in southern England and Wales since World War II and there are now almost 1000 hectares (2500 acres) of vines planted.

The history of the vine in Britain is of considerable interest and evidence suggests that it was the Romans, towards the end of the third century, who first brought the grape to England. Since that time little was written about vines, or anything else for that matter, until about the Norman conquest, although The Venerable Bede briefly refers to vines in the eighth century. The Domesday Book in 1085 mentions thirty-eight vineyards and, a hundred years later, William of Malmesbury said of Gloucestershire that: 'This country is planted more thickly with vineyards than any other in England, more plentiful in crops and

more pleasant in flavour. For the wines do not offend the mouth with sharpness since they do not yield to the French in sweetness.' A statement, no doubt, that a French patriot would find unbelievable. The tradition of vineyards was especially strong in the monasteries and was, subsequent to their dissolution, adopted by many enthusiastic farmers. But, by the time of Charles II, John Rose, his gardener, was already noting a decline in viticulture and states that it was due to the planting of vines in rich soil and the neglect of proper cultivation. A concept which will be encountered later in this book.

The decline of vineyards, due possibly to climatic changes and the dissolution of the monasteries—or to the availability of cheap wines from France — continued up to the present time. Nevertheless, in 1670, Sir Thomas Hanmer in Wales wrote a description of vine pruning which is effectively the cane pruning system described at some length in Part II.

Leave upon this head [stock] one chiefe or master branch ... let this master-branch be left halfe a yard or a yard long, according to the strength of the vine, and let it be the principal among the branches,

Figure 1.5 Distribution of vineyards in England and Wales.

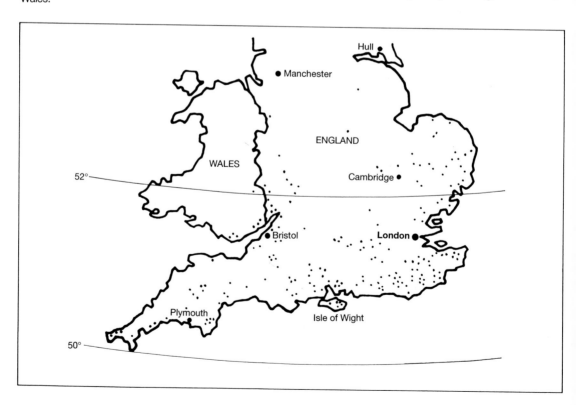

Denbies Wine Estate, Surrey.

and of the last yeare shoote, and the lowest of the other branches must be pruned very low or short, leaving only one eye or budd, or two at the most, and this short branch ... is to serve to send forth a master branch for the next yeare, cutting of[f] the master branch which was left the last yeare. This vine being thus pruned, cut away all the other branches except those two aforesaid, except you judge the vine so vigorous that it can nourish dowble as much.

Modern British viticulture owes much to a number of enthusiastic growers whose confidence in their ability to grow vines gave encouragement to others, and whose trials provided the background necessary to support a small industry.

Rainfall in Britain, generally, is not high but tends to be evenly distributed and very dry spells are a rarity. Lack of heat (Heat units 700-750°C, 1250-1350°F) and high humidity cause disease problems, inadequate sugars, and high acid in many years; nevertheless some very good light wines can be found and winemakers so far have had no problems in selling their product. The most favoured areas are the counties bordering the Thames Estuary—stretching as far as Bristol, the coast of Kent and along the south coast as far as Swanage and the Isle of Wight. Devon and Cornwall are warm, but rather too wet.

Of the grapes grown, Müller-Thurgau has been the most widely planted, although since it does not really adapt well to the local climate, its popularity is declining. The following grapes, in approximate order of ripening, are being used: Seiggerebe, Ortego, Optima, Madelaine Angevine, Reichensteiner, Bacchus, Huxelrebe, Schönburger, Kerner and Seyval blanc. A few red grapes are used, including Pinot noir which can make a pleasant light red wine when grown in appropriate positions.

Considering the small size of the industry there is a remarkable interest in the grape, and Britain must have more books on viticulture, in proportion to the hectares under cultivation, than any other country in the world.

CHAPTER 2

FACTORS CONTRIBUTING TO QUALITY IN SOME COOLER WINE DISTRICTS OUTSIDE EUROPE

California

The first move towards wine production in America was the planting, in Peru, of European vines by Cortez in 1524. Grape cuttings were established in parts of Mexico and Southern California by numerous missions for the production of sacramental wines during the seventeeth century. A Jesuit priest, Father Juan Ugarte, is credited, in the 1690s, with planting the first grapes in the cooler parts of the San Francisco district. The Mission grape, for a long time the most-planted grape in California, still survives in parts of the state.

By the late eighteenth century, vines could be found in most parts of Southern California, slowly extending further north as far as San Francisco. The wines produced varied from the port types around Los Angeles, to the 'hocks' and 'sauternes' of the Sonoma and Napa valleys, with 'sherries' being made in Sonoma and El Dorado counties. The 'clarets' of Napa and Sonoma were said to represent the better red wines made at that time.

The more extensive planting of quality grapes from the classical districts of Europe dates back to the middle of the nineteenth century. Large scale experimentation with the suitability of European varieties to Californian soils and climates was largely due to the efforts of Colonel Haraszthy. His experiments laid the foundation for the increases in production of quality table wines which are being seen in present times. Between Colonel Haraszthy's trial work and present day expansion, numerous set-backs had to be overcome; among them the collapse of California's gold rush with the subsequent lowering of wine prices in the 1860s and the economic decline of the early 1870s, followed by the

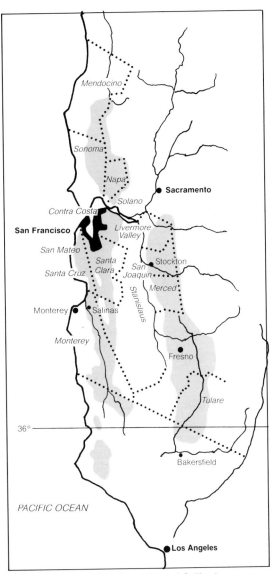

Figure 2.1 The major wine regions of California

27

Table 2.1 The major wines, varieties, soils and climates of the northern coastal districts of California

District	Wines	Varieties
Sonoma	White and red, some sparkling	Cabernet Sauvignon, Merlot, Gamay, Zinfandel, Pinot noir. Chardonnay, Riesling, (Johannisberger Riesling), Sauvignon blanc, Chenin blanc, Sémillon, Gewürztraminer and other lesser grapes
Napa	Mainly red, some white and sparkling	Petite Syrah (Durif), Cabernet Sauvignon, Merlot, Zinfandel, Pinot noir, Chardonnay, Chenin blanc, Sauvignon blanc and other lesser grapes
Livermore Valley	Mainly white, some red and sparkling	Pinot blanc, Sémillon, Grey Riesling (Trousseau), Sauvignon blanc, Chenin blanc, Riesling, Chardonnay, Grenache and other lesser grapes
Contra Costa	Red, white and some sparkling	Carignan, Mataro, Palomino, Pinot noir, Zinfandel, Refosco, Chardonnay, Chenin blanc, Sauvignon blanc and other lesser grapes
Santa Clara	Red, white and some sparkling	Petite Syrah, Carignan, Mataro, Cabernet Sauvignon, Merlot, Zinfandel, Chardonnay, Sémillon, Chenin blanc, Ruby Cabernet, French Colombard, Emerald Riesling, Pinot noir, Grenache and other lesser grapes
Monterey	White, red and sparkling	Chardonnay, Riesling, Sémillon, Chenin blanc, Pinot noir, Sauvignon blanc, Ruby Cabernet and other lesser grapes

appearance of *Phylloxera* in the latter part of the same decade.

The last decade of the nineteenth and the beginning of the twentieth century witnessed a large expansion of California's wine industry, so that grape-growing became a major industry. The quality of wine, however, improved very little during that time. Later came the disastrous, if brief, period of prohibition in the late 1920s and early 1930s.

The present output of wine stands at some 2400 million litres (627 million US gallons) from little over 400 000 hectares (990 000 acres) of vineyards. Roughly two-thirds of this total are table wines, the majority of which could be said to be slightly better than vin ordinaire and to resemble those of the lesser parts of Italy, France and Spain. The best of California's wines are equal in quality to the better European wines, and are invariably varietal wines produced from the non-irrigated vineyards in the cooler, climatic conditions of the north-coast counties.

Geographically, and from the standpoint of wine quality, the Californian wine regions can be divided into three major zones. The already-mentioned north coast region — including areas of Sonoma, Napa, Santa Clara, Livermore Valley, Contra Costa, Monterey and others — all producing fine quality, table wines. The other two zones, San-Joaquin or Central Valley, and

San Bernardino to the east of Los Angeles, provide a large volume of lesser quality wine.

It will be noticed that even the cooler areas of California tend to rate fairly high on LTI or Degree Days. In fact, they mostly come into Region II, and a number of grapes common in such regions are noticed in Table 2.1, eg Carignan, Palomino, Zinfandel, French Colombard, and Grenache. The relatively low latitudes, eg Napa 38° 24', means the growing season is long, but the summer temperatures are tempered by the coastal influence of the Californian current.

Wine legislation, though less restrictive than in Europe, has had beneficial effects on quality. California's State Board of Viticultural Commissioners, originally formed in the 1880s to assist growers in overcoming Phylloxera, has helped to stabilise wine production and, by sponsoring favourable state legislation, helped to establish basic quality standards for wine production. A more recent example of such legislation was the setting of the minimum amount of a single grape variety required in varietal wines, a stipulation which further improved the quality and reputation of local wines. The valuable research and teaching facilities at the University of California, Davis and Fresno State University can also be considered as an important quality factor in Californian

Soil Type and Position	Annual Rainfall	Heat Units In Growing Season	LTI
Varies greatly from the low fertility loams in Sonoma and Santa Rosa, to the more fertile, alluvial soils of the Russian River Valley, Mostly flats, some on gentle slopes in the Russian River valley.	700 mm (27 in) in Sonoma-Santa Rosa, to 1,350 mm (53 in) in Russian River Valley.	From 1200°C (2160°F) in Sonoma to 2000°C (3600 °F) in Russian River Valley	518
Fertile clay and silt loams in the south, gravel loams of better drainage and lower fertility in the north. Mostly flats, some on slopes.	500-650 mm (20-26 in)	1300-1450°C (2340-2610°F)	518
Gravel loams, often high in stone content and rich in limestone. Both flats and slopes are planted.	500-550 mm (20-22 in)	1250-1400°C (2250-2520°F)	542
Clay loams, sometimes gravelly of low to medium fertility. Flats and lower slopes are planted.	550-600 mm (22-24 in)	1400-1450°C (2520-2610°F)	528
Clay and gravel loams, varying in fertility from low to medium. Flats and lower slopes are planted.	550-600 mm (22-24 in)	1250-1300°C (2250-2340°F)	490
Gravelly well drained soils of low fertility, mostly over clay subsoils. Mainly flats.	450-500 mm (18-20in)	1200-1300°C (2160 2340°F)	410

Christian Brothers Winery, Napa Valley

wine production. Apart from tackling problems facing grapegrowers, students from these institutions also provide a new generation of highly-skilled growers and winemakers for the further improvement and expansion of the industry.

Present developments in California's wine industry include a more critical assessment of the climatic and soil conditions which prevail in wine areas in an effort to produce quality, varietal wines of distinctive types. This requires the introduction of selected clones of existing, or even new, varieties and the further development of vine training methods and winemaking techniques better suited to local conditions. There is a trend not to use names such as Chablis, Burgundy etc., thus giving more emphasis to local place-names and traditions.

The recent large-scale failure of the rootstock AxR-1 (ARG1) to control *Phylloxera* highlights the importance of using rootstocks that do not contain European vines (*Vitis vinifera*) in their parentage. It is estimated that two-thirds of Napa vineyards will need to be replanted in the mid 1990s due to the extensive use of this rootstock in the past. New plantings tend to be closer spaced, with less-vigorous rootstocks and use more modern canopy systems.

The Pacific North-west

As in other parts of the world, there is considerable interest in extending viticulture and winemaking to new and cooler areas. Oregon, Washington, and British Columbia have a potential for wine production which is now being exploited.

The Interior Valleys

In central and eastern Washington, grapes have been grown since the early part of the century when irrigation first opened up the semi-arid areas for horticulture. Predominant in development has been the American variety Concord which is used mainly for juice, jelly, and other non-alcoholic products. The area planted with Concord grapes increased dramatically in the early 1970s, while *vinifera* grapes for wine are now increasing in popularity. Some of these grapes are used for winemaking in Washington; other grapes are sent to Oregon and Canada to be made into wine.

In terms of area, the interior valleys of the

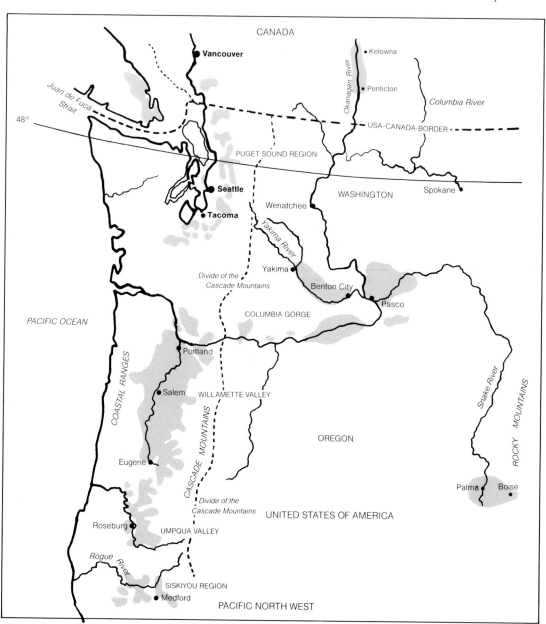

Figure 2.2 The Pacitic Northwest ot the USA and Canada

The Interior Valleys Ste. Michelle Winery, Washington

Pacific North-west are therefore the most important. Vineyards are centred in the Colombia Basin and the Yakima and Snake Rivers. Due to the Cascade Mountains, between this region and the Pacific, moisture-laden air is dried and produces an arid climate where irrigation is essential. The Rocky Mountains serve as a buffer to the frigid north-east Arctic air, so that in most years cold-susceptible *Vinifera* are not damaged.

Because of the northerly latitudes, summers are not long (150-180 days), but they are intense with 1100-1700°C days (2000-3000°F) being accumulated. This combination produces LTI values of 320-380, which suggests that Group IC grapes such as Cabernet Sauvignon,

Table 2.2: The major grape-growing districts of the interior valleys of the Pacific North-west

District	Wines	Main varieties	Degree Days	LTI
A. Boise/Parma, Idaho	White and red	Chardonnay, Riesling, Gewürztraminer, Pinot noir.	1390-1670°C (2500-3000°F)	380
B. Tri-Cities (e g Pasco), Washington	White and red	As above. plus Cabernet Sauvignon, Merlot, Chenin blanc, Sauvignon blanc, Sémillon.	1390-1670°C (2500-3000°F)	332
C. Wahluke, Washington	Red and white	Cabernet Sauvignon, Merlot, Chenin blanc, Sauvignon blanc, Muscat Canelli.	1670-1945°C (3000-3500°F)	330
D. Yakima Valley, Washington	White and some red	Riesling, Gewürztraminer, Chardonnay, Pinot noir.	1200-1500°C (2200-2700°F)	320
Umatilla, Oregon	White and some red	As above plus Müller-Thurgau, Sylvaner	1390-1670°C (2500-3000°F)	335
Plymouth/ Alderdale, Washington	White and red	Most varieties listed above, plus some Grenache	1390-1670°C (2500-3000°F)	335

Oregon: Hyland Vineyard, McMinnville

Sémillon, Sauvignon blanc could be grown. In fact, as described in the Introduction to this book, not all districts grow such varieties, because severe winter chilling is common in some locations and these varieties are more susceptible than most grapes of Groups IA and IB. The cool autumns preserve satisfactory acid and pH levels and quality is generally very good.

The area is served by the Washington State University Experimental Stations, such as the one at Prosser, which investigates grapes and other temperate fruit crops.

The major districts are shown in Table 2.2.

Western Oregon and Washington

To the west of the Cascade Mountains, the climate is more moderate, summer and winter temperatures are less extreme and rainfall is higher. Vineyards generally are not irrigated, although summers can sometimes be quite dry. Unfortunately, spring and autumn can often be moist and this can bring problems with poor set in spring and rots in the autumn. Climatic data are shown in Table 2.3.

The main grapes grown in the Puget Sound are those from Group IA. Group IB grapes are mostly used in the Willamette Valley, even

Table 2.3 Climatic data for western Oregon and Washington

District	Degree Days	LTI	Rainfall	Grape Group Predicted
Puget Sound	800-850°C (1440-1530°F)	136-145	980mm (39 in)	IA
Willamette Valley	1000-1200°C (1800-2250°F)	305	1060mm (42 in)	IC
Umpqua River Valley	1250-1360°C (2250-2480°F)	370	525mm (21 in)	IC
Rogue River Valley	1300-1440°C (2400-2600°F)	380	430-1524mm (17-60 in)	II

though the prediction is for Group IC. This is an excellent example of high rainfall lowering the capacity from a higher group to the one below. In the Umpqua River and Rogue River Valleys, groups IC and II will ripen, but for historical reasons growers have, in the past tended, to concentrate on grapes in group IB.

Varieties grown in the Puget Sound — a minor region—include Foch, Cascade, Müller Thurgau, Seyval, and Aurore. In northern Oregon, winemakers are particularly enthusiastic about Pinot noir, but excellent wines have also been made from Chardonnay, Riesling, Pinot gris, Gewürztraminer, and Cabernet Sauvignon.

Oregon is an exciting new district which has an enthusiastic and knowledgeable group of grapegrowers and winemakers who, together with research and advisory personnel from Oregon State University, are beginning to make an international reputation with their wines.

Canada

Canada is not generally regarded by the rest of the world as a grape-producing or wine-making country, yet there are currently nearly 12 000 hectares (30 000 acres) of land devoted to viticulture and indeed, grape production and wine consumption have increased dramatically until recently. This growth was encouraged by high import duties on wine from other countries. Now, however, with the moves to reduce tariffs with USA it is more difficult for Canadian wines to compete and production is declining.

Despite the hostile climate of much of Canada, two areas are well suited to viticulture: the Niagara Peninsula of Ontario and the Okanagan Valley of British Columbia. Viticulture in both these areas has, over the past fifty years, been supported by vigorous Government research programmes. New areas are continuously being considered, including the Annapolis Valley of Nova Scotia and the north shore of Lake Erie.

Grey Monk Winery overlooking Lake Okanagan

Historically, Canada has based its production on American grapes, chiefly Concord and Niagara, and much of the wine sold was fortified. In the last twenty years, however, the use of Franco-American hybrids and *Vitis vinifera* grapes have increased considerably and they are now predominant. As a consequence, since 1980, ninety per cent of the wine produced is table wine.

As in the rest of North America, the eastern areas tend to be wetter than the west. High rain and associated disease problems, together with soils which are often heavy clay and poorly drained, are the chief hazards associated with Ontario. Thus research, for example at the Vineland Station in Ontario, pays particular attention to breeding grapes with disease resistance and cold tolerance, as well as being concerned with wine quality characteristics. More recent plantings have been made in lighter, better-drained soils. In the Okanagan Valley, low rainfall is typical and sandy, well-drained soils are much more common. *Vinifera* grapes are subject to fewer disease problems, although cold winter temperatures can sometimes cause vine death.

The Eastern States

With the greatest area of vines under cultivation — producing some eighty per cent of the country's wine—and, so far, with the greatest reputation, California dominates the rapidly expanding American wine scene. Nevertheless, other American states and Canada have important wine industries and many, especially those in cooler parts, are expanding rapidly.

Historically, of course, it was in the eastern states that the first wines were made by the early colonists from the grapes growing in that area; and even before colonisation the land had been named 'Vinland' because of the profusion of new *Vitis* species which had been found there. These early wines began a tradition of wine production based on American species, in particular *Vitis labrusca*. They were different from European wines, and while connoisseurs tend to discount such wine and object to the so-called 'foxy' flavour, many feel that they have qualities which can be appreciated if the drinker is prepared to dissociate himself from the tastes and expectations of *vinifera* wines. Longfellow clearly was impressed when he wrote:

> *Very good in its way*
> *Is the Verzenay*
> *Or the Sillery, soft and creamy;*
> *But Catawba wine*
> *Has a taste more divine,*
> *More dulcet, delicious and dreamy. "*

Winemaking began as early as 1619 in Virginia and, possibly in the 1500's, the Spaniards planted vines in Florida. The first commercial vineyard in the USA was established in 1793 at Spring Mill, Pennsylvania.

It is now a developing industry which includes over 300 wineries—see map—and includes 32 states from New Hampshire to Minnesota, and from New Mexico to Florida. Four species of grapes contribute to the wine industry in the Eastern States, they are:

1. The imported *Vitis vinefera*, of which the most significant varieties are: Cabernet Sauvignon, Chardonnay, Riesling and Gewürztraminer.

2. Native *Vitis labrusca*, such as Concord, Niagara, Catawba, and Delaware.

3. The native *Vitis rotundifolia* grown in the deep south from Lousiana to North Carolina.

These produce the heavily-flavoured, sweet grapey wines of Muscadines, Scuppernong, Magnolia, and Carlos.

4. The Franco-American hybrids, of which the leaders are Seyval blanc, Vidal blanc, Ravat blanc and Aurora.

New York is the most important wine-producing state outside California, and the Finger Lakes district the most prestigious within the state. This area of long narrow lakes below Lake Ontario, south-east of Rochester, has been planted since the early nineteenth century, and as well as producing fortified and table wines, it makes almost half of the sparkling wine in the United States. Most have borrowed European designations for their wine — 'Sherry', 'Port', 'Sauternes', 'Burgundy', 'Champagne', etc. Some are blended with Californian wine but, increasingly, the better wines take their names from the grapes in the vineyard. There are several major wineries, of which perhaps that of Dr Konstantin Frank — 'Vinifera Wine Cellars' — is the best known, since the owner has been the leading light behind the move to classical grapes. His wines and those of other dedicated growers have made people aware that, with proper care and attention, and the assistance of modern technology, good *vinifera* wines can be produced in areas previously considered too wet and humid.

New York has another claim to fame in the world of grapes and wine. In various research institutes, such as the New York Agricultural Experiment Station, Geneva, much important research has been accomplished. The 'Geneva Double Curtain' training system developed by Dr Shaulis is known throughout the world for its high productivity, and our understanding of the way grapes grow has been significantly advanced by Dr Charlotte Pratt, Dr Robert Pool and many other viticulturists and oenologists whose research has proved to be so valuable.

Warm, humid summer weather characterises many eastern areas and the native vines which evolved under such conditions survive where, in the past, the European species have succumbed to fungus and the scourge of *Phylloxera*. The modern grower uses American rootstocks to fight the latter and a whole new armoury of sprays to control the fungi. With new technology in the winery, reasonable wine can be made even in those years where fungus cannot be adequately controlled.

Degree Days and LTIs for most vineyard

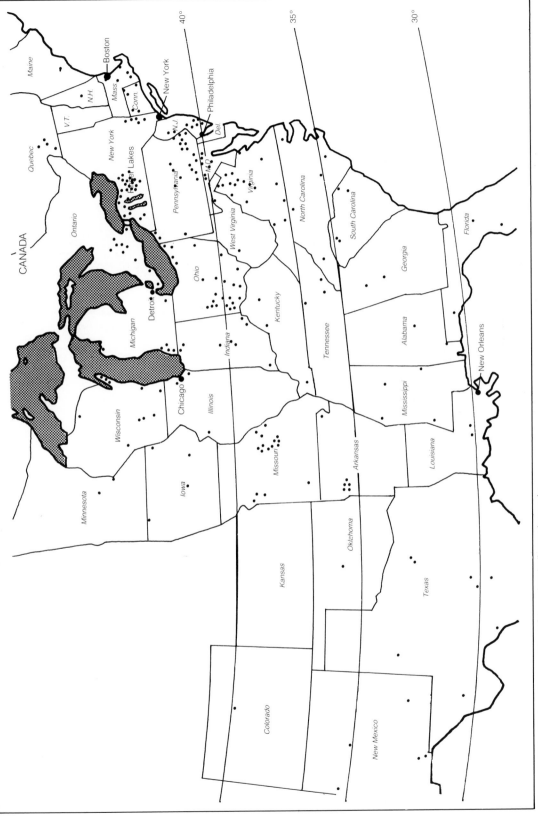

Figure 2.3 The Eastern States of the USA — dots indicate vineyards

areas are high and would be above the figures expected in a cool viticultural area. Nevertheless, the vinifera grapes which are grown tend to be in groups IA, IB and IC. This is due to two factors — firstly, the high rainfall in many districts and secondly, the severe winter chilling that is experienced, especially in the north.

Map and details included in this section are from *Eastern wine: a profile Vinifera Winegrowers Journal*, Spring 1983: 38-9.

Vineyard and golf course being successfully run as a single operation

Australia

The first reported plantings of grapes in Australia were made by Captain Arthur Phillip on a site which is now part of the Sydney Botanical Gardens. This was in 1788, but although the vines survived for a number of years, the wine produced was of mediocre quality, probably due to the warm, humid climate near Sydney Harbour. Phillip was not discouraged and three years later, planted 1.2 hectares (3 acres) at Paramatta, west of Sydney. This was more successful and encouraged other colonists to plant grapes. By 1803 viticulture must have been well established as a part of mixed cropping by numerous settlers, since an issue of the Sydney Gazette at the time included detailed instructions on how to plant and tend a vineyard.

By 1827 the vintage totalled an annual 91,000 litres (24 000 US gallons) and was selling well on the Sydney market.

In 1824, a young botanist, James Busby, came to Sydney from Scotland to supervise an orphan boys school. Attached to the school was a small vineyard and by tending this, Busby's enthusiasm for grapes was kindled. In 1830 he returned to Britain and while there took the opportunity to study European viticulture and oenology. At the request of the British Government, he took back to Australia 20 000 cuttings and established a vineyard in the Hunter Valley, which by 1857 had grown to 120 hectares (290 acres). It was largely due to his enthusiasm and persistence that the wine industry developed so effectively in those early years and today he is often referred to as the father of the Australian wine industry.

Following the initial success of grape-growing in New South Wales, another area emerged as a quality wine producer. Victorian viticulture began around Melbourne, Lilydale and the Yarra Valley. The initial 40 hectares (100 acres) in 1848 expanded in less than twenty years to some 1200 hectares (3000 acres). Growth in the established areas, plus those of Bendigo, Hawthorn, South Yarra, Rutherglen

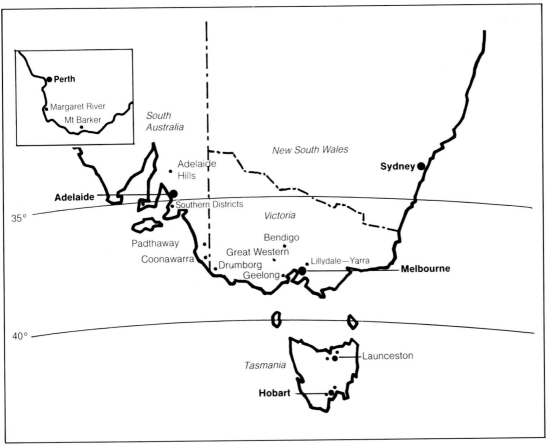

Figure 2.4 The cooler wine regions of Australia

Vineyard in the Adelaide Hills

and Great Western were responsible for this increase.

The prolific growth of Australian viticulture in the 1840s and 1850s was reflected in the third major grape-growing area to be planted. South Australia, though established late, was to become in future years the largest wine producer.

The first vineyards of South Australia were planted in the Barossa Valley, the Southern Vales, parts of Adelaide, Coonawarra, Clare and Langhorne Creek. From the 300 hectares (740 acres) planted in 1856, the area under vines increased to over 2440 hectares (6000 acres) in 1870 and, as the vineyard areas of Victoria declined, the importance of the South Australian wines increased. By the turn of the century, the once great Victorian vineyards were being devastated by *Phylloxera*, and the vineyards of South Australia—under the protection of rigid quarantine measures (1899) prospered still more.

The early part of the twentieth century in Australia witnessed some further expansion of viticulture in the irrigated areas of the Murray and Murrumbidgee rivers in South Australia, in New South Wales and to a lesser extent in Victoria.

During this time mainly fortified wines were

produced, though an increase in the production of table wines, both red and white, is significant from 1946 onwards when table wines accounted for some 40 per cent of the total volume.

The 1950s witnessed a boom—particularly in table wines — as a result of improved technology, increased affluence, widespread European immigration and the establishment of numerous wine promotion clubs and societies .

Wine popularity increased during the 1960s and early 1970s when table-wine production and consumption increased manyfold to the current level of 20.5 litres per head per year (1990)

Currently 70 per cent of wines in Australia are sold in soft packs (wine boxes) or flagons and much of this wine is produced in warmer, irrigated districts where mechanisation and economies of scale are used. Quality wines are still mainly sold in traditional 750 ml bottles, and many of the best are produced in smaller 'boutique' wineries in the cooler districts.

Such cool climates include the following districts:

South Australia: Adelaide Hills,
                Coonawarra, Padthaway/
                Keppoch.

Victoria: Great Western, Yarra Valley,
    Pyrenees, Geelong, Drumborg,
    Bendigo, Avoca, Macedon.
Western Australia: Margaret River,
    Mt. Barker.
Tasmania: Launceston, Hobart.
The grape varieties which produce the finest quality in such districts are:
Red: Cabernet Sauvignon, Cabernet Franc,
    Malbec, Merlot, Shiraz, Pinot noir.
White: Chardonnay, Sémillon, Sauvignon
    blanc, Gewürztraminer.
Australian grapegrowers and winemakers benefit by some excellent research stations, members of which have gained international reputations for their work. The Australian Wine Research Institute at Adelaide has achieved a great deal by raising the standard of Australian wines. Next door is the Waite Institute where valuable viticulture research has been done. The Division of Horticultural Research of the Commonwealth Scientific and Industrial Research Organisation (C.S.I.R.O.) stationed at Merbein has specialised in clonal selection and vine breeding, mechanical harvesting, training, etc.

Two Colleges offer training in viticulture and oenology. They are Roseworthy Agricultural College, north of Adelaide, and Riverina College of Advanced Education at Wagga Wagga in southern New South Wales. They are now respectively part of the University of Adelaide and Charles Sturt University.

Table 2.4 The major wines, varieties, soils and cool climates of Australia

District	Wines	Varieties	Soil Types and Position	Annual Rainfall	Heat Units	LTI
Adelaide hills (S.A)	Red and white	Riesling, Chardonnay, Sauvignon blanc, Shiraz, Cabernet Sauvignon, Malbec, Sémillon, Chenin blanc, Pinot noir, Gewürztraminer.	Varies from schist rocks on Trial Hill to more common sandy loams over gravel clay sub-soils near Springston and Keyneton. Gentle to steep slopes mostly above 400 m.	550-770 mm (22-28 in)	1150-1200°C (2070-2250°F)	550
Coonawarra (S A)	Red and white	Shiraz, Cabernet Sauvignon, Malbec, Merlot, Riesling, Chardonnay, Pinot noir, Sauvignon blanc.	Red clay loams over red-brown clay and limestone, sometimes with high stone content. Some poorly-drained sites in south-east section. Flats often with high water table.	600 mm (24 in)	1250°C (2250°F)	530
Padthaway— Keppoch (S.A.)	White and red	Riesling, Chardonnay, Sauvignon blanc, Gewürztraminer, Crouchen, Sémillon, Sylvaner, Chenin blanc, Muscadelle, Müller Thurgau, Cabernet Sauvignon, Malbec, Pinot noir.	Brown grey sandy loams over red-brown clay and limestone sub-soils, sometimes with high stone content. Mostly well-drained but some poorly-drained sites in the south-east. Flats with high water tables.	550 mm (22 in)	1250°C (2250°F)	550
Mt Barker (W.A.)	Red and white	Cabernet Sauvignon, Shiraz, Pinot noir, Malbec, Merlot, Riesling, Chardonnay, Sauvignon blanc, Sémillon, Gewürztraminer.	Well-drained alluvial gravel loams over clay, rich in limestone. Low fertility. Flat sites or gentle slopes. Salinity of irrigation water can be a problem.	900 mm (35 in)	1400°C (2500°F)	492
Margaret River (W.A.)	Red and white	Cabernet Sauvignon, Merlot, Malbec, Cabernet Franc, Shiraz, Pinot noir, Zinfandel, Sémillon, Riesling, Chardonnay, Sauvignon blanc, Chenin blanc	Gravel and sandy loams over clay sub-soils. Well-drained but with high water-holding capacity of sub-soil. Gentle slopes, salt spray can be a problem.	900- 1000 mm (35-38 in)	1540°C (2750°F)	577

Table 2.4 Continued

District	Wines	Varieties	Soil Types and Position	Annual Rainfall	Heat Units	LTI
Yarra Valley (Vic)	Red and white, some sparkling	Cabernet Sauvignon, Pinot noir, Merlot, Shiraz, Chardonnay, Riesling, Gewürztraminer, Sémillon, Sauvignon blanc.	Volcanic red basalt or grey loams over clay sub-soils. Some drainage problems on flats, good drainage on slopes. Some excessive vigour problems on fertile soils.	900-l000 mm (35-38 in)	1150°C (2070°F)	454
Great Western (Vic)	White, red and sparkling	Chasselas, Ondenc, Pinot meunier, Mataro, Chardonnay, Riesling, Gewürztraminer, Cabernet Sauvignon, Merlot, Malbec, Pinot noir, Shiraz.	Sandy-clay loams over clay on flats, gravel and sandy loams over clay on slopes Flats and gentle slopes. Frosts in spring can be a problem.	525 mm (21 in)	1500°C (2700°F)	504
Drumborg (Vic)	White, red and sparkling	Riesling,Gewürztraminer, Sylvaner, Chasselas, Pinot noir, Cabernet Sauvignon, Muscadelle, Chardonnay, Ondenc.	Volcanic, red-brown loams over gravel-clay sub soils, often rich in limestone Mostly flat.	750 mm (30 in)	1200°C (2160°F)	446
Bendigo (Vic)	White and red	Chardonnay, Riesling, Sémillon, Sauvignon blanc, Gewürztraminer, Shiraz, Chenin blanc, Cabernet Sauvignon, Pinot noir	Variable sandy gravel, volcanic basalt or clay loams mostly over clay sub-soils. Frost in spring a problem in flat vineyards.	500-550 mm (20-22 in)	1250-1300°C (2250-2340°F)	493
Geelong(Vic)	Red and white	Shiraz, Cabernet Sauvignon, Pinot noir, Gewürztraminer, Riesling, Chardonnay.	Reddish clay limestones or well-drained deep volcanic basalts over clay or limestone sub-soils. Gentle slopes or flats.	530 mm (21 in)	1200°C (2160°F)	423
Macedon (Vic)	Red and white	Cabernet Sauvignon, Shiraz, Malbec, Merlot, Cabernet Franc, Riesling, Sauvignon blanc, Chardonnay, Sémillon, Gewürztraminer	Coarse granite soils and sandy loams over clay sub-soils. Best sites are on gentle slopes at higher elevations.	700-750 mm (28-30 in)	1200°C (2160°F)	474
Moonambel - Pyrenees (Vic)	Red, white and sparkling	Cabernet Sauvignon, Merlot, Malbec, Shiraz, Cabernet Franc, Sauvignon blanc, Chardonnay, Riesling, Trebbiano, Chenin blanc.	Well-drained gravel-quartz loams over clay sub-soils, in parts rich in limestone. Gentle slopes.	600 mm (24 in)	1350-1400°C (2430-2520°F)	490
Avoca-Pyrenees (Vic)	Red and white, sparkling	Trebbiano, Chardonnay, Sémillon, Cabernet Sauvignon, Shiraz, Malbec	Gravel and sandy loams over clay sub-soils Flats and gentle slopes.	550-600 mm (22-24 in)	1400°C (2520°F)	510
Launceston and Hobart (Tas)	Red and white	Cabernet Sauvignon, Pinot noir, Riesling, Gewürztraminer, Chardonnay, Sauvignon blanc, Müller Thurgau	Isolated areas with a wide variation of soils from well-drained sandy-gravel loams to clay soils, both over clay sub-soils. Mostly slopes of varying steepness.	750 -800 mm (30-32 in)	1000-1150°C (1800-2070°F)	342

New Zealand

Grapes for winemaking were first planted in New Zealand in the early nineteenth century by French settlers and religious missions. At first only the European, *Vitis vinifera* varieties were grown, and the wines produced were said to be of good quality.

The progress of viticulture was slow in many parts of the North Island, since the climate was not suitable for grape-growing and making wines in the European tradition. Owing to high rainfall and high humidity on many of the original sites around Auckland, grapes were often lost to disease, and those surviving had little chance of producing quality wine. The methods adopted by growers from the European wine regions proved to be of little help in the local warm and humid conditions.

In February 1895 the government brought in the services of the Italian, Romeo Bragato, who was at that time the viticultural advisor to the Victorian government. His report, issued in September of the same year, could have been an excellent blue print for the New Zealand

Figure 2.5 The major wine regions of New Zealand

Table 2.5 The major wines, varieties, soils and climates of New Zealand

District	Wines	Varieties	Soil Type and Position	Annual Rainfall	Heat Units in Growing Season	LTI
Northland	Red and white table wines, some fortified	Cabernet Sauvignon, Chasselas, Palomino, Pinotage, Müller-Thurgau.	Shallow clay soils over sandy-clay subsoils Flats and mild slopes.	1600 mm (63 in)	1300-1400 °C (2340-2520 °F)	450
Auckland (Henderson and Kumeu)	White and red table wines, some fortified and sparkling	Müller-Thurgau, Pinotage, Chasselas, Cabernet Sauvignon, Merlot, Gewürztraminer, Pinot noir, (including Gamay de Beaujolais), Sémillon and Sauvignon blanc.	Shallow clays over hard silty-clay subsoils or sandy loams. Mainly flats.	1500 mm (59 in)	1300-1350 °C (2340-2430 °F)	440
Waikato	White and red	Chasselas, Müller-Thurgau, Chenin blanc, Cabernet Sauvignon, Pinotage, and others	Heavy loams over clay subsoils. Flats and mild slopes.	1100-1200 mm (43-47 in)	1250-1300 °C (2250-2340 °F)	414
Gisborne	White and red table wines, fortified and sparkling	Palomino, Müller-Thurgau, Chasselas, Pinotage, Chenin blanc, Chardonnay, Cabernet Sauvignon, Pinot noir, Merlot, Gewürztraminer	Fertile, alluvial loams over sandy or volcanic subsoils. Flats.	1000-1050 mm (39-41 in)	1250-1300 °C (2250-2340 °F)	394
Hawkes Bay	White and red table wines, some fortified and sparkling	MüllerThurgau, Chasselas, Pinot gris, Cabernet Sauvignon, Chardonnay, Merlot, Sauvignon blanc, Riesling	Clay loams of medium to high fertility over gravelly or volcanic subsoils. Flats.	750-800 mm (30 32 in)	1200-1250 °C (2160 2250 °F)	384
Wairarapa	White and red	Pinot noir, Chardonay, Pinot gris, Merlot, Riesling	Deep stony and silt loams over gravel	1050	1080-1150°C (1944-2070 °F)	332
Nelson	White and red	Müller-Thurgau, Pinot noir, Chardonnay, Gewürztraminer, Riesling, Refosco, Cabernet Sauvignon, Sylvaner.	Clay loams over hard clay subsoils. Slopes	1000-1250 mm (39-49 in)	1050-1100 °C (1890-1980 °F)	320
Marlborough	White, red, and sparkling	Müller-Thurgau, Cabernet Sauvignon, Pinotage, Chardonnay, Riesling, Gewürztraminer, Sauvignon blanc, Pinot noir, Muscat Dr. Hogg Merlot, Pinot meunier.	Silty-alluvial loams over gravelly subsoils. In parts compacted silt or clay pans of various thickness and depth are found. Flats.	650-750 mm (26-30 in)	1150-1250 °C (2070-2250 °F)	327
Canterbury	White and red	Gewürztraminer, Müller-Thurgau, Riesling, Pinot gris, Pinot blanc Pinot noir, Chardonay, Cabernet Sauvignon, and Sauvignon blanc.	Alluvial silt loams over gravel subsoils in the central parts. Chalky loam soils often rich in limestone in the northern part. Gentle slopes.	600-750 mm (24-30 in)	900-1100 °C (1620-1980 °F)	277
Central Otago	White and red	Riesling, Pinot noir, Chardonay, Sauvignon blanc, Pinot gris, Gewürztraminer.	Silt loams with mica and schists. Moderate to steep slopes	400-450 mm (15-18in)	850-1000°C (1530-1890 °F)	260

Vineyard in Marlborough, photograph by courtesy of Montana Wines Ltd

wine industry but, regrettably, it remained largely ignored.

He returned to Victoria, but was back in New Zealand in 1902 as Viticulturist and Head of the Viticultural Division of the Department of Agriculture. He expanded the research station at Te Kauwhata and produced an excellent handbook in 1906. This notwithstanding, his efforts were largely ignored by the government of the day and by 1909 he had become disillusioned. He resigned and returned to Italy.

Phylloxera was already present in New Zealand by that time and Bragato had recommended the grafting of *vinifera* grapes on American rootstocks. Unfortunately many growers ignored his advice and instead planted Franco-American hybrids and *labrusca* grapes which, in a short time, became predominant.

With the introduction of hybrid grapes, the hopes for quality wine production were lost until the 1960s. The strong prohibitionist movement during the first two decades of this century did little to encourage growers and winemakers to plan for a long-term future. Dual-purpose grape types, for wine and table consumption, produced high yields of fruit which proved to some extent resistant to humidity at ripeness. Ordinary table and fortified

wines were made from Siebel, Baco, and Albany Surprise grapes. The better types included Palomino, Chasselas, and more recently, Pinotage, Cabernet and Müller-Thurgau. To avoid potential spoilage, even the better vineyards were often harvested prematurely and the widespread virus infection of local vineyards prevented grapes from reaching good maturation levels needed for quality wine. To compensate for excess acidity and low sugar, winemakers had to resort to the use of water and to sugaring their wine heavily and, at best, an ordinary wine quality resulted. At one stage the shortage of wine grapes caused winemakers to buy all grapes that were offered to them — often with little regard to their ripeness and quality.

The 1950s witnessed a dramatic increase in the demand for wine—mainly sparkling and fortified. Increased vineyard planting in Auckland and Hawkes Bay areas resulted, and although the average wine made was of fair quality only, several fine table wines were also made, thus showing the potential for future years.

The 1970s were the decade of rapid expansion and growth in the New Zealand wine industry, and several large companies joined

—and even surpassed in size—the until-then leaders, McWilliams of Napier and Corbans of Auckland. Amongst these winemakers, the spectacular growth of Montana, Villa Maria, Penfolds, Delegats, Cooks and Nobilos must be noted. There are now two major groups, Montana and Corbans, several medium-to-large groups including Villa Maria and Nobilos and many medium-to-small wineries.

Next to the greatly increased volumes of wine made, a major change has occurred in the quality and the types produced, and in their acceptance by a rapidly expanding market. For example, previously the most planted grapes, in area, were Baco 22A, Albany Surprise, Palomino, Siebel 5445 and Müller-Thurgau. By 1981, only Müller-Thurgau remained in the top ten.

Major varieties by area planted in 1993 were: Chardonay (21%), Müller-Thurgau (18%), Sauvignon blanc (14%), Cabernet Sauvignon (8%), Pinot noir, Riesling, Muscats, Chenin blanc, Merlot, Gewürztraminer.

The total area planted in grapes reached 6800 hectares in 1993, producing in average vintages, close to 60 million litres of wine. The main quality factors influencing the current production of wine can be summarized as follows:

• The use of early-ripening grapes, well suited to conditions found in the three principal districts; Gisborne, Hawkes Bay, and Marlborough.
• Rapid modernisation of wine cellars and the widespread use of quality, table wine production methods.
• Improved spray materials giving better disease control, especially in wetter parts of the country, thus allowing the grower to pick riper grapes.
• Introduction of improved planting material via regional Vine Improvement Groups.
• Expansion of the wine production in the cooler districts of Hawkes Bay, Wairarapa and the South Island.
• The introduction of legislation (1983) controlling the quality of wines produced, and the rapid expansion of research into both grape-growing and winemaking.
• The recent export drive by the New Zealand wine industry which received favourable responses in such diverse markets as the United Kingdom, Scandinavia, and North America.

By the mid-eighties some of the euphoria of the seventies had disappeared and overproduction of certain varieties such as Palomino, Müller-Thurgau, Chenin blanc and Gewürztraminer had occurred. New plantings have concentrated on premium varieties geared to the export market and have provided dramatic changes in the industry. Amongst the white wines Sauvignon blanc, Chardonnay and Riesling are of recognised international standard. The better quality red wines are based on Cabernet Sauvignon and Merlot, with Pinot noir proving successful in southern regions. Despite the introduction of wine sales to supermarkets, the domestic wine production is static at 16 litres per head. From the 1995 vintage a regional denomination system will guarantee the wine's origin and will encourage the recognition of distinctively regional characteristics.

The wetter districts of the North Island have degree days and LTI which put their potential ripening ability into Region II. Few such grapes are, in fact, grown since these districts have maximum autumn and winter rainfall and the useful ripening period is reduced. Because of New Zealand's island character, summer temperatures never become excessively hot and in late summer and autumn — when ripening occurs—temperatures are cool to warm so that, with good vineyard management, well-balanced musts can be obtained.

SOURCES OF INFORMATION AND FURTHER READING

Adams, L.D. 1984. *The Wines of America*, 3rd ed. McGraw-Hill Co., New York.

Ambrosi, H. (translated Gavin, B.) 1976, *German Wine Atlas*, Ceres-Verlag Rudolf— August Oetker K.G., Bielefeld.

Amerine, M.A., Singleton, V.L. 1977. *Wine, An Introduction for Americans*, 2nd ed. Univ. of Calif. Press, Berkeley and Los Angeles.

Anon. 1980. *Osterreichisches Statistisches Zentralamt*, Vienna.

de Blij, H.J. 1983. *Wine — A Geographic Appreciation*, Rowman and Allanheld, New Jersey.

Bragato, R. 1906. *Report on the Prospects of Viticulture in New Zealand*, N.Z. Govt. Printer, Wellington.

Carosso, V.P. 1951. *The Californian Wine Industry 1830-1895*, Univ. of Calif. Press, Berkeley.

Cattel, H., Lee Stauffer, H. 1979. *The Wines of the East. 1. The Viniferas, 2. The Hybrids*, Land H. Photojournalism, Lancaster, Pa.

Chroman, N. 1973. *The Treasury of American Wines*, Routledge—Crown, New York.

Coombe, B.G., Dry, P.R. 1988. *Viticulture, Volume 1, Resources in Australia*, Winetitles, Adelaide.

Cooper, M. 1993. *Wines of New Zealand*, Houghton and Stanton, Auckland.

Deutsche Weininformation. 1971. *Deutscher Weinatlas*, Frankfurt.

Eggenberger, W., Peyer, E. 1982. *Weinbuch 1982*, Fachverlag Sweizer Wirteverband, Zurich.

Halliday, J. 1985. *The Australian Wine Compendium*, Angus and Robertson.

Halliday J. 1991 *Wine Atlas of Australia & New Zealand*, Collins, Auckland.

Hyams, E. 1949. *The Grape Vine in England*, Bodley Head, London.

Johnson, H. 1985. *The World Atlas of Wine*, Mitchell Beazley, London.

Johnson, H. 1983. *Modern Encyclopedia of Wine*, Simon and Schuster, New York.

Kuhnholtz, L.G. 1963. *La Genèse des Appellations de'Origine des Vins*, Buguet et Coumptour Press, Macon.

Lichine, A. 1967. *Encyclopaedia of Wines and Spirits*, Cassell, London.

Loeb, O.W., Prittie, T. 1972. *Moselle*, Faber and Faber, London.

Mayne, R. 1986. *The Great Australian Wine Book*, Reeds, Sydney.

Morton, L.T. 1985. *An Illustrated Guide to Viticulture East of the Rockies*, Cornell Univ. Press, Ithaca.

Norton, R.A. 1975. *Growing of Grapes for Table and Wine in the Puget Sound Region*, Co-op Ext. Service, Washington State Univ. E.M. 3910.

Ordish, G. 1977. *Vineyards in England and Wales*, Faber and Faber, London.

Pearkes, G. 1984. *Vinegrowing in Britain*, J.M. Dent and Sons, London.

Poupon, P., Forgeot, P. 1972. *The Wines of Burgundy*, 4th Ed. French Univ. Press, Paris.

Poupon, P., Jaquelin, L. 1960. *Vignes et Vins de France*, Flamarion Press, Paris.

Proesler, H. 1974. *The First Traces of Viticulture on the Rhine*, Allgemeine Deutsche Weinfachzeitung, Neustadt.

Schreiner, J. 1984. *The World of Canadian Wine*, Douglas and McIntyre, Vancouver/ Toronto.

Schuster, D.F. 1965-1987. *Notes on the Vineyards of France, Germany, South Africa, Australia and New Zealand,* (unpublished).

Simon, A. 1957. *The Noble Grapes and Great Wines of France,* McGraw-Hill, New York.

Skelton, S. 1989. *The Vineyards of England,* S.P. & L. Skelton, Kent.

Stewart, K. 1985. *The New Zealanders' Guide to Wine,* Hodder and Stoughton.

Thorpy, F. 1970. *Wine in New Zealand,* Collins, Auckland.

Vivez, J. 1956. *Les Appellations de 'Origine,* Librarie Polytechnique, C. H. Beranger, Paris.

Ward, J. 1984. *The Complete Book of Vine Growing in the British Isles,* Faber and Faber, London.

Weaver, R.J. 1976. *Grape Growing,* John Wiley and Sons, New York.

Winkler, A.J., Cook, J.A., Kliewer, W.M., Lider, L.A. 1974. *General Viticulture,* Univ. of Calif. Press, Berkeley

PART II

GRAPE GROWING

INTRODUCTION

Grapes, which belong to the genus *Vitis*, are distributed throughout North America, the Caribbean, Asia and the Mediterranean. The North American species are by far the most numerous and include varieties* which are of value as table grapes, juice and rootstocks, but none that produce wine of any note. The European grape, *Vitis vinifera*, comes originally from the Caucasus and, since its introduction to Europe, has been hybridised with the native vines *Vitis silvestris* and *Vitis alba*. It is common for European grape varieties, which account for the majority of the world's wine—and all of that which can be termed classical—to be referred to simply as *Vitis vinifera*.

The historical accident that introduced the root-aphid, *Phylloxera*, to Europe—with the subsequent destruction of thousands of hectares of vineyards—caused growers and scientists to look for ways and means to save the industry. Two methods were considered the most promising. The first was to cross European grapes with species from America which had a natural resistance to *Phylloxera*. The crosses produced the Franco-American hybrids which were hoped to have the quality of the former and the resistance of the latter. Regrettably, these aims were not generally achieved. These hybrids have been grown commercially in New Zealand, in the eastern states of America and in parts of Europe; their area is declining and in some countries their use for wine is prohibited. The second method is universal in regions where *Phylloxera* is endemic and, since the aphid does not damage the upper part of the vine, involves the grafting of *Vitis vinifera* to resistant rootstocks of American origin. The result is entirely satisfactory if the right rootstock is chosen for the right variety of grape and the appropriate soil and climate, although it makes propagation more difficult and expensive.

Hybrids between *Vitis vinifera* and American *Vitis* species are very productive and have good disease and cold resistance. Very often, however, they make only mediocre wine.

*Varieties of grapes, or any other cultivated plant, are more correctly called 'cultivars'. We have generally retained the term variety due to common usage and the practice of naming wines made from any one variety as 'varietal'.

CHAPTER 3
THE GRAPE AND ITS GROWTH

Root growth

The roots of vines have the capacity to penetrate deep into the subsoil and, while the majority may be in the top 1-1.5 m (3-5ft), deeper roots may move down 4 m (13 ft) or more providing there are no impediments to growth. Most nutrients and water will be taken up by roots in the topsoil, but under conditions of moisture-stress the vine will depend more and more upon deeper roots to supply water for growth and development. Vines are renowned for their ability to survive in arid areas but, if their roots cannot penetrate into the lower soil profile, the vine's capacity to thrive in such areas will be little better than that of any other plant.

Practical measures to encourage root penetration are, firstly, subsoiling or ripping, and secondly, deep ploughing Subsoiling is used to break up hard pans which often occur below the topsoil and restrict root growth in the subsoil. It is done either through the whole vineyard or simply along the lines where grapes will be planted. Deep ploughing, which may bury the top soil 60-90 cm (2-3ft), encourages roots to move down to the more fertile topsoil and discourages surface rooting. At the same time it moves soil to the surface which is free of weed seeds, and weed control for a number of years is reduced.

Shoot growth

In spring, after dormancy has been broken and temperatures increase, buds swell and shoots emerge. These shoots are soft and tender and susceptible to frost, so consequently, in cooler areas, frost-protection measures or frost-free sites may need to be considered. It is fortunate that bud burst is relatively late in these areas and at a time when most frosts have finished.

Growth of the shoot is dependent not only on suitable temperatures and adequate sunshine, water, and nutrients, but also on complex physiological factors within the plant. Very important among these factors are growth regulators or hormones. These, like animal hormones, exist within the tissues at very low concentrations and regulate the growth of the plant or animal. Dormancy, for example, probably results from buds and seeds having high levels of inhibitory hormones (inhibitors) which restrict growth even if temperatures are high. This is of advantage to the plant for it prevents buds bursting in the autumn, or during warm spells in winter, and then being killed by subsequent frosts. As spring approaches inhibitors decline in concentration and promoters increase, thus buds will then grow as the temperature rises.

The growth of shoots continues to be under the control of hormones during the season. Being a vine, the rate of growth is faster than most other deciduous plants—2-3 cm (1in) a day at the most rapid stage — and continues for a longer period into the season. In apples, for example, shoots make a flush of growth of 30-40 cm (10-12 in) in the first six weeks after bud burst and then many make little or no further growth until next year. In grapes, shoots, and laterals from those shoots, continue to grow until berries are about three-quarters full-size and, in warm and humid climates, even later.

Growth of shoots produces leaves which, by the process of photosynthesis in sunlight, manufacture carbohydrates such as sugar.

50

Figure 3.1 *Top:* Young shoot four-five weeks after bud burst. *Bottom:* Young lateral and dormant bud in axil of leaf.

Carbohydrates are used by the developing shoots, leaves, buds, roots and berries. If the use of carbohydrates exceeds production, reserves are used and this is the case in spring before the leaf area has reached a critical size. There are times when it can be advantageous to remove developing shoot tips to allow the materials which they consume to be temporarily moved elsewhere. At the time of flowering, pinching the tips, spraying with a growth retardant which reduces vegetative growth, or girdling can improve set. If vegetative growth is still active at berry ripening, Removing the top 10-25cm off the shoot tip ('topping') may advance maturity.

Formation of flower buds

Buds are formed in the axils of leaves as the shoots develop. Two parts of the bud can be discerned; one remains dormant until the next spring, while the other may develop to form a short or long lateral shoot (see Figure 3.1)

Within the dormant bud there are three growing points. The primary one is central and is the most fruitful. The other two, termed secondary and tertiary, are less fruitful. In spring the primary bud will normally grow and the other two remain dormant. Occasionally the secondary will also grow — this is dependent on a number of factors, for example:

1. Some varieties commonly produce multiple shoots, Gewürztraminer is a good example.
2. Heavy pruning will induce more multiple shoots to grow; the grower is then left with a high percentage of shoots which have a reduced number of flowers. Sometimes these are broken off by the grower.
3. Late spring frosts may kill the primary shoot, after which secondary and even tertiary shoots may form. The amount of crop gained will depend on the fruitfulness of the secondary shoot — the tertiary one seldom produces flowers. Normally the second crop will produce from 25 to 50 per cent the yield of the primary one.

Leaf and inflorescence* initials in the primary bud begin to develop in early summer, almost a year before flowering, and within six to eight weeks development is complete. These initials will remain more or less in this form until the following spring. To improve the development of inflorescence initials, which will determine next season's crop, it is necessary to consider growth which occurred during the previous season.

---

* The structure which carries the flowers is called an inflorescence; later these stalks plus the berries are called the bunch or cluster. Although the inflorescence is formed in the previous season, the flowers do not form till spring.

A critical leaf area is needed to promote initiation and any trimming of shoots that removes large areas of mature leaves could be detrimental. Flower initiation is also promoted by high light intensity and the amount of light actually falling on the developing bud and adjacent leaves is critical. If the foliage is very thick then it follows that light penetration to the leaf axils will be reduced and, consequently, flower initiation will suffer. The grower therefore should manipulate his/her canopy so that there are adequate leaves for initiation (and fruit ripening), yet the canopy should not be too dense so that it prevents light reaching the bud sites. (Some 8-10 sq cm of exposed leaf area are required to ripen 1 gram of fruit.) Heavy applications of fertiliser—especially nitrogen — and excessive water will induce poor initiation; this is partly a consequence of excessive shading.

Temperatures, also, have a marked bearing on floral initiation. Strangely, these temper-atures have their effect in the first few weeks, from the time the node subtending the bud changes from the shoot apex to ten nodes back on the shoot. This is at a time well before any microscopic evidence of flower initials can be seen. Optimum temperature for floral initia-tion shown by Aus-tralian work was found to be around 30°C (86°F). Once again, heavy shading will be deleterious because temperatures in the shade are cooler.

The fruitfulness of buds varies according to the position of the bud on the shoot. Generally, buds formed close to the base are less fruitful (i.e.will form fewer bunches next spring), than those six to seven nodes higher. Those closer to the tip of the cane also become less fruitful, as shown in Figure 3.2.

Cane pruning, which ensures that more of the buds in the middle cane positions are used, is more popular in cool climates. In hotter and drier climates, where bud initiation is usually better, spur pruning, using more buds near the base, is common.

Flower set

When buds burst in spring, leaves appear and flowers do not open until six to eight weeks later. Over the rather inconspicuous flower a cap is positioned and when anthers are ready to be released the cap is shed and the pollen from the anthers falls on the stigma. Viticulturists often refer to this stage as *capfall, anthesis* or *full bloom*. Pollination is assisted by light winds and is not, as with many plants, dependent on insects such as bees: it does occur more readily, however, in warm dry weather. Two to three days later fertilisation is complete and in another two to three days those flowers not fertilised begin to drop off—a stage known as *shatter*. Set refers to the flowers which have been pollinated and fertilised and which remain on the bunch.

Inadequate fertilisation (sometimes called coulure) can have three consequences; firstly, as mentioned above, flowers fall off shortly after capfall; secondly, small green berries form which may drop a little later, although occasionally, on some varieties, they may remain until harvest. The third possibility is that the berries may apparently develop nor-

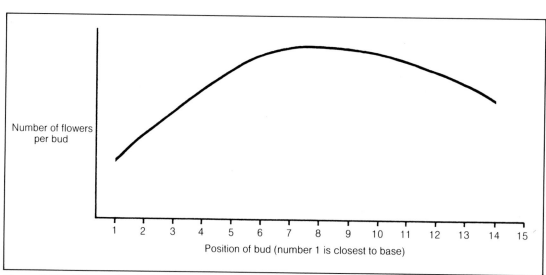

Figure 3.2 Fruitfulness of buds along a cane

mally, but never attain full size—bunches containing normally-set large berries and small berries are called 'hen and chickens' or 'millerandage'. Examination of chickens will show that they may be seedless or have fewer and or smaller seeds than normal. It is a common phenomenon in many plants that fruit size is directly related to the number of seeds.

There are three ways that a grower can use to increase set. These are listed below:

1. The grower tops the vine a few days before cap fall. Since the tips of the shoots are competing with the flowers, it is believed that topping temporarily gives the flowers a good supply of food materials and/or hormones.
2. Vines can be girdled about a week before capfall; this is done on the trunk or on the canes, and parallel bladed knives are available for this purpose. Two parallel cuts are made 3-4 mm ($^{2}/_{16}$in) apart about 10 cm below the head or 10 cm along the cane, and the bark between them is removed. Cuts should not damage the wood and, especially in cool climates, it must not be too wide. The method is only recommended on low-yielding varieties, especially when hen-and-chickens is common. Our own research has shown yield increases of 100 per cent on a poor clone of Chardonnay and 50 per cent on Gewürztraminer. There was no fall-off in effectiveness even if the vines were treated in three consecutive years. It is believed that the girdle temporarily stops the flow of photosynthates from the leaves to the roots and gives the flowers a competitive advantage at a critical period.
3. Chlormequat (CCC or 'Cycocel') is a growth retardant and, when sprayed on vines, it reduces shoot growth and increases set. Its effectiveness is equivalent to girdling and it works on the same varieties. It is not recommended for normal or heavy yielding grapes and the use of the correct concentration is essential. Concentrations of 500-700 ppm active ingredient applied about one week before anthesis are normally satisfactory. CCC will also reduce the amount of subsequent topping and trimming. Its use is prohibited in some countries.

A physiological disorder called Early Bunch-stem Necrosis (EBSN) can cause an apparent reduction in flower set. This is described in more detail in Chapter 9.

Second set or second crop is the term given to bunches which form on laterals from the main shoots. Because they are produced later in the season, they tend not to ripen in cool climates and can be a problem, especially when mechanical harvesting is used. The presence of unripe berries reduces the quality of the juice. It is common on many varieties, for example Pinot noir and Pinot gris, and can be accentuated by topping and CCC.

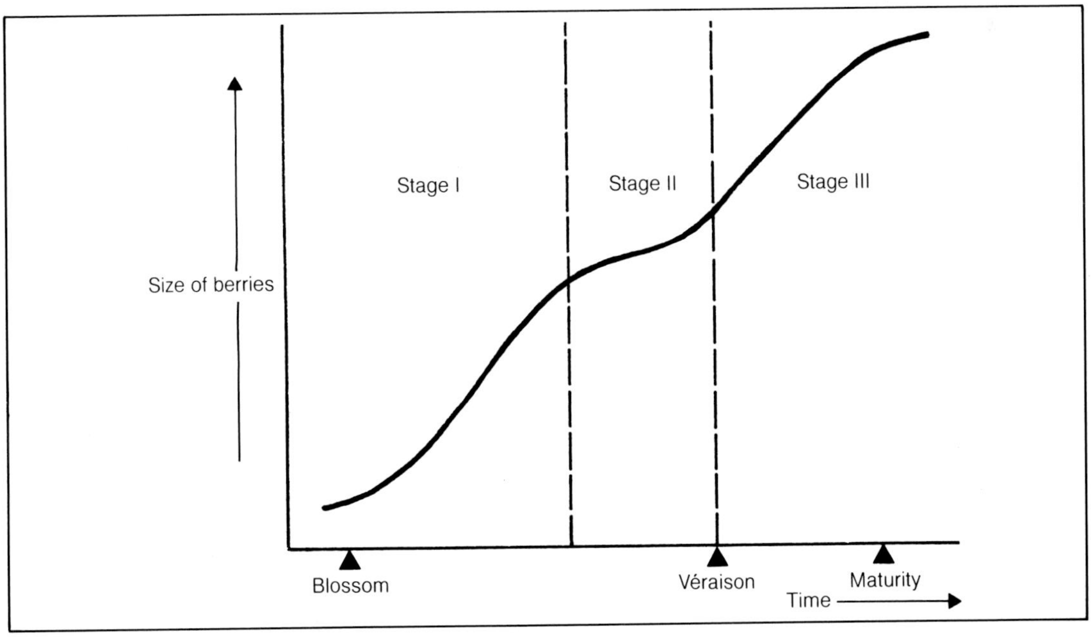

Figure 3.3 Growth stages of a grape berry

Growers often remove them by hand after veraison when they can be readily distinguished from the first set.

Fruit growth

The growth of fruit after flowering and set occurs in three stages. A period of rapid growth is followed by a slow growth period and culminates in another rapid growth phase prior to maturity, as shown in the Figure 3.3.

There has been a considerable amount of research into the understanding of the growth patterns of fruit. Again, it is believed that growth regulators or hormones play an important role in development. During Stage I hormones from the developing seeds are vital to the continued growth and retention of the berry on the bunch. Some varieties can develop without seeds, presumably by some genetic change in developmental physiology, but in cool climates, and for wine production, seedless varieties are not suitable. In Stage II, berry growth is slow but seeds develop rapidly to their maturity, which occurs at the end of this stage. At the end of Stage II, a sudden and dramatic change in the course of development occurs which is called *véraison*. At this stage growth accelerates again; the berry begins to soften, glucose and fructose increase, acidity decreases, chlorophyll is lost and the colour develops in red and black varieties.

At *véraison*, acid levels are high, but after this the levels are reduced by dilution, caused by an inflow of water into the berries, and by the conversion of acids to salts. Tartaric acid drops first and then remains more or less constant. Malic acid drops at a constant rate and may even disappear, so that tartaric acid is the only one left.

In hot climates, malic acid is lost very rapidly, whereas in cool climates levels remain high for longer. Malic acid, which is the common acid in apples, being prominent in cool-climate grapes, often tends to give the wine a taste reminiscent of apples. Wines of cooler areas are characterised by higher acidity levels; they are more complex and tend to need a longer period of maturation to achieve balance — in cold years excessive acidity can be a problem. In warmer climates loss of acidity, together with a general lack of complexity in the wine, poses different but still serious problems.

Colour changes in Stage III are particularly noticeable in red grapes, and again development is temperature-dependent in many varieties. Pigmentation of skins is greater in cooler temperatures, and in areas with greater temperature contrasts between day and night. Dry conditions, adequate, but not excessive, leaf area and moderate, rather than heavy cropping, promote the development of colour, whilst excessive nitrogen decreases it.

Apart from acids, sugars and colour constituents, there are other compounds which give grapes and wines their characteristic flavour and aroma. These include tannins and the many flavour and aroma constituents that develop, particularly in the later stages of ripening. This is one reason why the final weeks, or even days, are so significant in the development of wine quality. Rain, disease or frost can lead to premature picking and the loss of a potentially outstanding wine.

CHAPTER 4
THE CLIMATIC REQUIREMENTS FOR SUCCESSFUL GRAPE PRODUCTION

In the Introduction, the basic aspects of climate were considered. Although it is usually easier to grow grapes in warmer districts, it was shown that cool climates have some specific advantages for the production of quality wine. In this chapter, consideration of climate will have a more practical slant and site selection, frost problems, rainfall, and humidity will also be discussed.

Mesoclimates

Three convienient terms are used when discussing climates: *macroclimates, mesoclimates* and *microclimates*. A *macroclimate* is the general climate of an area. When this is modified by goegraphic features or by man-made factors such as shelterbelts or trees the term *mesoclimates* is used. An even smaller sub-division is a *microclimate*. Thus, within a vine, the inside of a canopy will be more shaded and often cooler than the outside.

As indicated, a number of factors can produce a *mesoclimate*. A site or a slope facing the sun, for instance, will be warmer than a flat site or one on the other side of the hill. If the hill gives shelter from prevailing cold winds it will be warmer still and if, for example, the soil contains many stones—and weeds are suppressed — further heat will be gained. Such sites can gather up to 200 extra °C days in a season and can make an apparently unsuitable district more attractive.

Figure 4.1 shows different temperature zones in a hypothetical, geographic area.

(a) A warm site catching more sun owing to the lie of the land—it misses late spring and early autumn frosts, since the cold air will drain to low-lying areas.

(b) The advantages of (a) will be counteracted by the cold which comes with altitude.

(c) A cold site: although it may miss frosts in spring and autumn, it will accumulate much less heat in summer due to exposure to wind and a poor angle to sun.

(d) Very cold and the area most susceptible to frost—cold air from surrounding districts will drain into this area.

(e) Still frosty but less than (d). Some shelter

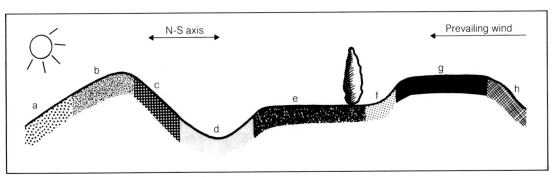

Figure 4.1 Mesoclimates on hypothetical sites

Wind Protection.
Wheat sown between the rows provides wind protection
for young vines. Returns from the wheat covered the cost
of drilling and maintenance

from wind may be obtained by the hill behind.
(f) The trees densely planted at the base of the
hill prevent cold air draining away and a potentially frost-free site has been lost.
(g) Less frost than (e), but a prevailing cold
wind and altitude may prevent the accumulation of warm air in summer.
(h) Cold, like (c) above.

Trees and shelter belts can be used to create
mesoclimates and will allow warm air to accumulate even on days with a cool wind. Care must
be taken to ensure they do not prevent cool air
draining away as in (f) and in Figure 6.1 in
Chapter 6.

Frosts

Temperatures of -3 °C (25 °F) or less in the
zone where leaves and inflorescences are positioned can be damaging. In spring young shoots
will be damaged or, if frosts are severe or last for
a long period of the night, they may be killed.
Secondary or tertiary shoots will then grow and,
although further fruit could be formed, yield will
be reduced and grapes may not ripen before late
autumn. Exposed flower clusters will be damaged
by -2 °C (28 °F) or less. Fortunately the buds of
grapes burst relatively late—about three weeks
after most stone fruit (peaches, apricots, plums
etc.), and normally at about the same time as
apples and pears. In autumn, frosts will cause the

leaves of the vines to drop and more severe frosts
can damage fruit, lack of leaves will reduce the
ripening capacity of the remaining fruit.

Frost-free mesoclimates in areas where spring
or autumn frosts are common are therefore
mostvaluable and, if not available, then some
frost-protection measures may be needed. A
grower should examine the degree of frost incidence in spring and autumn and decide whether
the risk of occasional damage is compensated
for by the cost of frost-protection measures.
Water-sprinkling is the easiest method of providing frost protection. During the hours when
the temperature is at danger-level, water is sprin-
kled over the vines—sprinkling begins when
the temperature drops to 1°C (34 °F) —and,
although it will freeze on the leaves and berries,
the constant addition of new water will prevent
the temperature from falling below 0°C (32 °F).
Such an installation can be automated and, of
course, the system will also deliver water for
irrigation during the summer. The rate at which
it is delivered over a given area is most important and it is recommended that local advice be
sought before installing a system.

Temperature can also be lifted by the use of
oil-burning frost-pots, or by using large fans to
mix the warm air above the frost zone with the
cold air immediately below it, so that the overall
temperature is raised to above the critical level.

Other methods which can be used to lessen
the damage caused by frost are:
1 . *Cultivation.* The temperature over bare soil
will be warmer than over weedy or grassed areas.
The soil will absorb more heat in the day and re-
radiate it at night, especially if the soil is moist
(but not excessively wet) and compacted rather
than loose.
2. *Height of pruning.* Normally the closer to the
ground the buds are positioned, the greater will
be the frost danger. Canes laid down at 30 cm
(12 in) will be more likely to be damaged than
canes at 1 metre (3ft).
3. *Pruning.* Pruning may be delayed until the
buds on the tips of the canes start to burst; those
buds closer to the head are much slower to open
and, after the removal of the top growth by
normal pruning, will then begin to develop.
Being a week or so later in opening will mean
they might miss early frosts, since closed buds
are much more resistant. The problem with
this is that pruning all the vines in such a
restricted time will be difficult to achieve: an
alternative is shown in Figure 4.2.
4. *Varieties with late bud burst.* The list in Table
8.1 gives the difference in bud burst of some

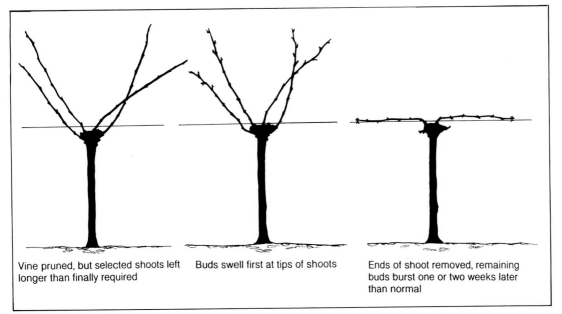

| Vine pruned, but selected shoots left longer than finally required | Buds swell first at tips of shoots | Ends of shoot removed, remaining buds burst one or two weeks later than normal |

Figure 4.2 Late pruning of vines for frost control

common cool climate grapes. The grower may find small differences in his/her own district and personal observation over the flowering period can be very valuable. Naturally, the later the buds burst, the more chance there is of escaping from damage.

Frost in the winter can cause problems to the vine in areas where temperatures are extremely low. This factor has been discussed in the Introduction.

Water requirements

Most viticultural regions of the world have a rainfall below 700-800 mm (27-31 in). Low rainfall, especially during berry maturation, is beneficial in reducing disease incidence and enabling grapes to be picked at the most appropriate time. Some well-known grape-producing areas such as the Rhine and Mosel in Germany have an autumn minimum with adequate rain in summer to enable vines to be produced without irrigation. Other areas are less fortunate and have an autumn maximum: some North Island districts of New Zealand and areas in Oregon west of the Cascades have such a pattern.

While, in many traditional areas of the wine-growing world, irrigation of vines is discouraged, growers in newer districts should not automatically assume that this applies in their own district. There are reasons why irrigation is often frowned upon and these are mostly concerned with quality. The two major reasons are:

1. Irrigation, especially by overhead sprinklers may encourage the spread of disease.
2. Over-generous application of water will induce excessive growth of shoots, thus inducing shading of lower and inner leaves, plus bunches; this can have an effect of reducing wine quality. Shading of shoots and buds will also reduce the number of bunches which will form next year.

As a general rule water stress in the vineyard should be avoided prior to véraison since it will reduce yield and possibly lower quality. Symptoms of stress include: stunted vine growth, abnormally short internodes, poor berry set, yellowing of leaves and defoliation, premature lignification. Later, uneven and imperfect ripening of berries will be noted. As will be described later, irrigation after véraison is often discouraged.

Evapotranspiration

There are two factors to consider when deciding whether to irrigate. These are: firstly, the water-holding capacity of the soil, and secondly, the potential water deficit occurring in the summer. The former will be considered in the next chapter. The water deficit is the difference between the useful rainfall occurring during the growing season and the water loss which occurs by evaporation from the surface of the soil, and by transpiration from grape leaves: together they are called evapotranspiration. Water loss is proportional to the loss of water from the surface of a volume of water in the vineyard, and, in meteorological stations, this evaluation is measured in a device called a Pan Evaporimeter, Class A. Often daily water loss is recorded in local newspapers.

If, for example, 5 mm ($^2/_{10}$ in) evaporation is recorded in one day, then the loss from the vineyard will be a proportion of that. The proportion is called the *crop factor*.

Calculating the crop factor

If the surface of the soil is completely covered by leaves—for example, if grass or weeds are growing between the vine rows — then the crop factor will be almost 1.0—in other words, water loss will be effectively equal to the evaporation from a Class A Pan. If, however, a considerable surface is bare soil, the loss will be much less since little water is lost from the surface of bare soil. It is difficult to be too specific about the crop factor but, as a guide, it is probably close to 0.5 if rows are 3m apart, and close to 1.0 if rows are 1m apart.

Deciding on irrigation

The determination of whether to install irrigation will be based on two factors, 1) the soil and its waterholding capacity, and 2) the difference between water gain and water loss, ie: rainfall—(evaporation x crop factor).

Thus, if the average rainfall for a month was 50 mm (2 in) and evaporation in the same month averaged 150 mm (6 in), and if the crop factor was 0.5, the balance for the month would be 50—(150 x 0.5), ie: 25 mm (2 - (6 x 0.5), ie: 1in) and, unless that 25 mm (1in) could be supplied by the soil, it can be expected that the plant would begin to show some stress. This discussion will be continued in the next chapter when soil reserves are discussed.

Other considerations

Vines can tolerate high rainfall and excess soil moisture while dormant in the winter as long as prolonged waterlogging does not occur. If supplied with an excess of water during the two months before harvesting, the vines will probably continue vigorous vegetative growth and the crop will most likely have only modest sugar and flavour intensity. Excessive rain can also cause berry splitting and fungus problems. Rain in warm temperature conditions is more damaging than rain in cool temperatures. In investigating a new district it can be quite revealing not only to discover the rainfall in the two months before vintage, but to find out whether it is predominantly warm or cold. The latter is much preferred.

Even in areas where water is normally adequate, occasions will arise when drought conditions are experienced. If irrigation is used in a very dry area, or in years when drought occurs in an area of normally adequate rainfall, it is considered best to irrigate before véraison and not after. Heavy irrigation to bring the soil to field capacity just before véraison will normally ensure adequate moisture for ripening whilst, at the same time, it will discourage excessive vegetative growth during berry maturation. The grower should not wait for signs of moisture stress to appear before applying water.

Two pieces of equipment can be used to assist the grower in determining appropriate irrigation: the *tensiometer* and the *neutron-probe*. Both are inserted into the soil and give a more direct reading of the water status. The former is cheaper and simpler, the latter is more expensive and is often operated by professionals who then advise the grower on appropriate irrigation scheduling. Details of the use of *tensiometers* and *neutron-probes* is best sought from local advisors.

CHAPTER 5

SOIL MANAGEMENT

The study of soils and their improvement is a complex one, but the general principles which apply to grape soils are similar to those for any other tree or bush. This chapter looks at the specific factors which need to be considered in preparing and maintaining the land for grapes.

General requirements

One of the first questions asked by prospective growers of grapes is: 'Don't grapes need special soils?' This idea is perhaps generated by wine enthusiasts who, in talking of wines from different districts, may emphasise the merits of one by arguing that the specific quality of the wine is due to the high chalk content of the soil or the deep gravelly loams. However, there are many other factors which affect quality: for example, the local mesoclimate, the variety or the blend of varieties, the time of picking, the training systems employed in that district, the nature of the yeasts used, the technique of winemaking adopted, or the level of skill of the winemakers. In fact, grapes grow in a wide range of soil-types, and variations in soils are among the many important factors affecting wine quality. Some general likes and dislikes of grape soils will be mentioned here and these will be amplified at the end of this chapter in the discussion on 'Terroirs'.

As a general rule heavy clays, shallow soils with impermeable pans below, those that are poorly drained, or soils with high levels of salts should be avoided. Deep soils, those with gravel or sand or with a moderate chalk content, are often ideal as can be seen from the soil condition tables in Part 1. High-fertility soils may produce excessive leaf growth and cause problems in spraying and disease control so that quality may suffer: lesser-fertility soils may reduce the crop yield but produce finer quality wine. Grapes, because of their deeply penetrating roots, will often grow in areas where other plants will not survive or be economic.

Chapter 8 discusses the soil preferences of certain varieties. These may not be firmly fixed rules, but experience has shown that the described varieties have produced good grapes and wines on such soils.

Fertility

If the soil has high fertility it would be unwise to add manures or fertilisers. The grower should have a soil analysis made before planting, and at intervals of two to three years, soil and leaf analyses will provide a continued check on nutrient levels. If deficiencies are found, specific fertilisers can be added to correct them. If a reasonable balance is indicated, it might be better not to fertilise unless poor growth occurs, at which time a general mixture can be added in spring and perhaps again in midsummer. Many growers will fertilise plants with a general fertiliser for the first two or three years after planting to get strong vigorous trunks and canes, and then add no more fertiliser unless specific problems are encountered. This is recommended as a good general rule in all but very rich or poor soils.

Cultivation

Most vineyards in the world are clean-cultivated. This means that no permanent grasses or clovers are allowed to grow, and the vineyard is cultivated at regular intervals to ensure bare soil beneath the vines and between the rows.

Clean cultivation has the following *advantages*:

1. Clean-cultivated soil loses less moisture than

soil covered by weeds or grasses. In dry areas this can help the grapes withstand drought in the summer without irrigation.

2. There is less humidity over bare ground than over grass or weeds, thus fungal rots are discouraged.

3. Bare soil absorbs more heat during the day and gives out more at night. Overall heat accumulation over the season is greater in a clean-cultivated vineyard than one with grass and, in spring or autumn, the danger of frost is reduced due to the slightly warmer night temperatures. More rapid air drainage over bare soil may also reduce frost on sloping ground by allowing cold air to drain away during the night.

4. Because grasses and weeds compete for nutrients, there are generally more nutrients available to the vine on soil that is clean-cultivated.

There are, however, some *disadvantages*:

1. It is always more pleasant to work in an orchard or vineyard covered with grass than in one covered with clods of earth.

2. Competing grasses may reduce surplus nutrients and water which might be an advantage in over-fertile or wet soils.

3. On steep slopes, heavy rain may wash considerable quantities of soil away; grass on the surface will significantly reduce this erosion. Grass will also aid rapid moisture penetration on flats in heavy rain.

4. The organic status of the soil is better under a mown rather than a cultivated vineyard.

Despite these disadvantages, the advantages are generally considered of more significance in cooler districts, and clean cultivation is generally recommended. In certain circumstances grassing the strip between adjacent rows may be contemplated perhaps initially on an experimental basis.

There is no need to over-cultivate—a few weeds will do little harm; excessive cultivation will destroy the soil structure and a continuous movement of tractors will compact the soil. It is advisable to cultivate to varying depths, preferably with different implements, so as to avoid the creation of an impacted boundary-layer.

Occasional deep, winter ploughing of clean-cultivated vineyards is used in many cool-climate districts. Deep ploughing is not carried out too close to the plants but covers the area up to and between the tractor wheels. Plough-ing opens up the ground to winter frosts which may improve its structure and kill vine pests, it may assist the distribution of fertilisers and fertility and also aid the penetration of winter rainfall into the subsoil. These factors, plus the reduction of compaction, assist root penetration and are valuable for all but steep slopes without terraces.

Weeds and grasses can be allowed to grow after harvest and the ground need not be cultivated again until just before bud burst. This will improve the soil structure and it may reduce erosion on sloping ground. If the weed growth is not too vigorous it could make the surface easier to work on, but it will be untidy, and if serious perennial weeds are present, it will assist their build-up. Excessive weed growth can make winter work unpleasant. Growers will need to assess for themselves the level of winter cultivation which best suits their conditions. Sheep are sometimes brought into vineyards in winter to reduce excessive growth of weeds.

The use of herbicides

Herbicides are being used increasingly in orchards and vineyards. Local advisers will be able to recommend the best and safest varieties which are currently available.

There are three basic types of herbicides:

1. Those that kill germinating weeds and young seedlings and, when applied to the soil, keep the ground weed-free for several months if not cultivated — examples, dichlobenil, simazine.

2. Those that kill existing growth but have limited or no residual effect — examples; paraquat and diquat.

3. More potent herbicides of the second type that may be needed for perennial and hard-to-kill weeds — examples; amitrole, asulam and glyphosate.

Combinations of '1' and '2' or '1' and '3' are the most common and are generally applied together in spring under the vines. These keep weeds controlled in the row for several months and considerably reduce the amount of hand-hoeing that is required. Complete weed control by chemicals between and within rows has been tried in orchards and vineyards, and they can be very satisfactory.

Extreme care must be adopted when using herbicides: some are very toxic to man, some will kill vines if wrongly applied, and toxic residues can build up in soils if they are applied too often and at higher-than-recommended concentrations. If used wisely, they can be a valuable aid to grape growers.

The relationship between soil and water

Different soils have different capacities to absorb and retain water, and an understanding of the basic principles behind the relationship can be very helpful to the grower or potential grower. It is important, first of all, to be familiar with a number of terms, these are:

- Saturation
  After rainfall or irrigation, water fills the spaces between particles of soil. If all the air is replaced by water the soil is said to be saturated. In fact, however, few soils can become totally saturated although most approach the saturation point.
- Field capacity
  A period (24 hours) of rapid drainage after saturation will cause the soil to retain a volume of water called the field capacity.
- Evapotranspiration
  More water may be lost by slow drainage after field capacity has been attained, but most of the further loss is due to evaporation from the surface of the soil and transpiration from the leaves of the vine — evapotranspiration.
- Permanent wilting point
  After a period of evapotranspiration plants may begin to wilt and if they are unable fully to recover, the soil water content is said to have reached the point of permanent wilting.
- Available water
  This is said to be the difference between the water content at field capacity and that at the permanent wilting point.
- Water table
  The level to which underground water may rise (see Figure 5.1). If it comes too close to the surface it may kill roots. A deep water table which can be reached by roots may act as a supplementary water source. Most viticultural soils do not have an accessible water table.

The amount of available water in a soil varies according to its texture and structure. This is clearly indicated in Table 5.1.

A soil with good structure will hold more water than a soil with poor structure. Thus, a vineyard newly planted in an area which previously was in grass or clover will be likely to have better structure than a vineyard which has been cultivated for a number of years with the addition of no organic matter. The available moisture would then be at the high end of the range given in Table 5.1.

Figure 5.1 Water table

If it is assumed that soils are at field capacity in the spring—which is normal in most cool-climate viticultural areas—and if we know the annual rainfall for the growing season (6-7 months), these data can give a rough guide to the total available water. For example, a siltloam, 90 cm (35 in) in depth, over a hard pan, will provide 182 x .90 = 163 mm (2.18 x $^{35}/_{12}$ = 6.3 in) of water; add to this the expected rainfall — say 300 mm (12 in) — and total water available is 463 mm (18 in). In cool climates, below about 1500°C (2700°F) days, a mature vineyard will need between 400-600 mm (16-24 in) of water

Table 5.1 Available moisture in soils

Soil texture	Available moisture—millimetres of available water per metre depth of soil (mm/m)—or inches per foot.			
	Range		Most Common	
	mm/m	in/ft	mm/m	in/ft
Sand	0-66	0-0.8	42	0.5
Sandy Loam	93-126	1.1-1.5	110	1.3
Silt Loam	177-186	2.1-2.2	182	2.2
Clay Loam	160-177	1.9-2.1	169	2.0

From: Goh. K.M.1986. Introduction to Garden Soils, Fertilizers and Water. Bascands Limited, Christchurch

during the growing season to sustain adequate but not excessive growth (see Chapter 4). Thus, this particular soil in this locality would probably need additional irrigation in many years to provide good growth and yields.

The critical point, if such calculations are to be used, is the depth of soil which can be fully exploited by the roots. Roots can penetrate to a depth of 4-8 m (12-25 ft) in sandy loams, sands or gravels, but it could be many years before there are sufficient roots at this depth to be able to extract all the available water. In clay soils, however, penetration is much less and may be less than 1m (3ft). An estimation must therefore be made of the soil depth which is available to the vine. It must be realised that any calculations made of water needs and water availability are at best approximations, but they can be a useful general guide in a new area if no other information is available on the response of vines in that district to irrigation.

Terroirs

The French term *Terroir* is used to describe the soil and its immediate environment (the site). The effect *terroir* has on the growth of the vine and the quality of the wine made from the grapes produced is the basis for French vineyard classifications. Good *terroirs* provide most or all of the following features:

1. Sites which warm rapidly in the spring to encourage early root and shoot growth.
2. Soils with sufficient nutrients and water to allow early unimpeded growth of the shoots until flowering.
3. Sites where some factor, such as moderate water or nutrient stress, reduces and eventually stops shoot growth after flowering, yet does not seriously curtail berry development.
4. Soils which seldom or never develop serious water or nutrient deficiency during the growing season.
5. Sites which have the capacity to drain away or otherwise shed excessive moisture after heavy rain, particularly close to harvest.

A good *terroir* has a buffering effect on climatic events so that, in wet years, excessive growth does not occur and in dry years severe stress is avoided. Wine from good *terroirs* varies in quality less than it does from poor *terroirs*.

The following *soils* have produced good *terroirs*:

1. Free-draining soils, often containing stones which warm up quickly in spring. Such soils should have sufficient fine particles such as silt and/or be sufficiently deep to make adequate water available over the season.

2. Shallower soils often with small stones or limestone particles to assist drainage and a higher proportion of fine particles — silt and clay — to hold adequate moisture.
3. Chalk soils which give good root penetration and drainage yet are not over-fertile. Chalk will sometimes act to allow water to percolate upwards to the roots from moisture below.
4. Clay soils which are not too heavy and will allow slow percolation of water. The ideal clay is one which expands with rainfall to prevent excessive absorption of moisture and so does not encourage new shoot growth and/or berry splitting after heavy rain.

The following *sites* are often associated with good *terroirs*:

1. Sloping sites with an inclination to the sun (S or SW in the northern hemisphere, N or NW in the south). These give good light exposure and improve water and air drainage (see Figure 4.1).
2. Sites sheltered from cooling winds.

Although, as we have noted, good *terroirs* will often reduce vigour, the productivity is not necessarily poor. Good light exposure and an appropriate leaf/fruit ratio improves plant efficiency and good *terroirs* are in fact capable of ripening a greater amount of fruit to a greater level of physiological ripeness than poor ones.

New World vineyard owners have paid less attention to *terroirs* than have those of the Old World. This is probably to be regretted. Nevertheless there is one factor which makes the New World grower rather less dependent on *terroir* than the Old. This is irrigation. Under EC Quality Wine Regulations irrigation is permitted only under 'exceptional climatic circumstances' and this means the soil, together with natural rainfall, must act to provide an appropriate water regime for the vines all year. The New World grower who is allowed to irrigate can effectively use a wider range of potentially-drier soils in more arid climates.

The French, while conceding irrigation has its benefits and may even, in some years, improve wine quality, believe it has inherent dangers. The main one is that it will lead to excessive use by growers who may be more interested in quantity than quality — this in turn may lower the reputation of the district. Despite these and other arguments for and against irrigation, the principles and concepts of *terroir* are sound and should be considered by growers when choosing a site for a vineyard. Appropriate modification to the concept will be made for sites where irrigation is to be used.

CHAPTER 6

ESTABLISHING A VINEYARD AND PROPAGATING VINES

Choosing the Site

The types of soil and climate preferred by vines have been described. The position of the land is also significant: not only should it be relatively close to markets or wineries, but there are, in addition, a number of aspects affecting growth and production which need to be considered.

Shape of plot

Square or rectangular plots are the easiest to handle and enable the maximum use of land. Irregular land causes problems since more 'dead' or unplantable areas are encountered, rows are of different lengths, tractor operation is more difficult, and workers have further to walk to reach all parts of the vineyard.

Slopes

Very steep land should be avoided. It only has merit in very special areas where the value of the crop compensates for the extra effort needed in cultivation, pruning, picking etc. Light slopes, while not so easy to manage as flat land, have certain advantages in cool areas. Additionally, sloping land is less liable to have drainage problems than flat land. Planting on slopes will be considered in detail later in this chapter.

Preparing the Site

Preliminary cultivation

As already mentioned, deep soil is preferred, and if there is any hint of an impermeable pan below the surface, ripping or sub-soiling should be done. The best time for this is towards the end of summer, while the ground is relatively dry down to the pan. If it is done when the ground is wet it can do more harm than good. Deep ploughing of 30-60 cm (12-23 in) has also been suggested in order to bury the fertile top-soil and encourage the roots to move to the deeper soil layers.

It is vitally important to rid the ground of any perennial weeds. Local advisory officers will describe the best methods of control and these must be carefully followed. Perennial weeds around the base of vines are much more difficult to overcome once the vineyard is planted and can entail hours of unnecessary work.

Establishing shelter

Once on the trellis, vines are probably no more sensitive to wind than other horticultural crops. Nevertheless, areas with prevailing winds will benefit from shelter belts as physical damage and persistent movement of vines will restrict vegetative growth, especially during the establishment years—sheltered areas are also warmer. Protection is maintained, on the leeward side, for approximately ten times the height of the shelter belt.

Tall, narrow, deciduous trees like Lombardy Poplars make good shelter belts as they take up a minimum of space and provide fewer nesting sites for birds. Being leafless in winter, deciduous trees have the additional advantage of permitting air movement which helps the soil to dry out in early spring. Shelter can be said to be more effective if a moderate amount of air movement takes place: if a solid, impermeable shelter is used, turbulence occurs on the leeward side which diminishes the sheltering effect.

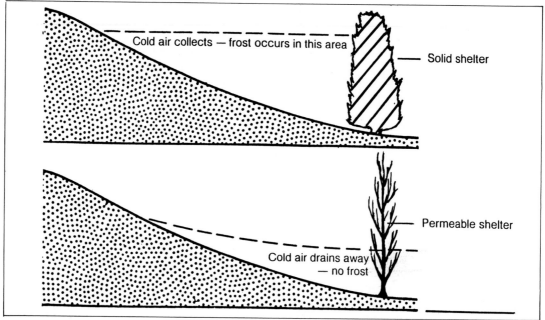

Figure 6.1 Shelter variations

Shelter, however, should not be used unless necessary. It wastes space, shades vines, competes for water and nutrients and may, as previously noted, prevent the movement of cold air away from a vineyard on a slope, see Figure 6.1. Bird predation is often considerably enhanced when local tree numbers are increased.

Planting Vines

Planting may take place at any time when leaves are off the vine. Land, however, is normally more workable in autumn or spring and these are the optimum times. If grafted plants are used, the graft should always be above the soil surface; nevertheless it is desirable that planting should be as deep as possible to encourage deep rooting and to help

the plant survive dry conditions in the first and subsequent summers — two buds above the surface are all that is needed. Sometimes, after planting, the soil is mounded up over the plants; this will protect the buds from severe winter frosts and provide a couple of weeks' extra protection in the spring from frosts which can kill the young shoots. Once the growth is showing above the mound the soil is taken away.

In wet areas, where *Phylloxera* is not a problem, cuttings are sometimes planted straight into the vineyard. The moisture and humidity allow the plant to survive the critical first weeks before roots are established. In drier areas, planting cuttings through black plastic can be considered. Plastic is laid down the rows over moist soil and cuttings are kept in a cool shed or cool store until three to four weeks after

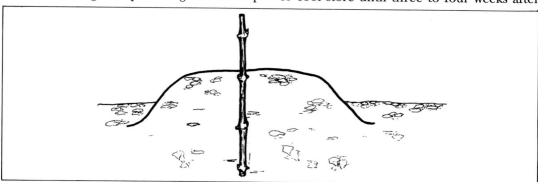

Figure 6.2 Planting grape cuttings directly in the vineyard through black plastic

normal bud burst. They are then pushed through the plastic into their final position. By this stage the soil is warm under the plastic and roots are quickly established in the warm moist environment.

The plastic will last two to three years; it will encourage the rapid growth of vines and discourage weeds. Permanent posts for the trellis are driven through the plastic immediately, or after one or two seasons if stakes are provided for each plant.

Spare cuttings are placed in a nursery. It is advisable to allow for 25 per cent loss and the spare cuttings are used to replace any cuttings in the vineyard that die. After planting in the nursery a pre-emergence herbicide sprayed over the soil will reduce the need for hoeing.

Vineyard Layout

Generally, growers prefer to have rows in a north-south direction. This ensures that both sides of the vines receive similar amounts of sunshine and ripening will occur at about the same rate. If there is a prevailing wind in the area it is desirable to have rows at right angles to the wind, the first row lifts the air over the subsequent rows: if the rows are in the same direction as the wind, the wind will travel down the row and do more damage. Vineyards planted on the slope usually have the rows going up and down — this, it is believed, allows cold air to drain away on frosty nights and warm air to drift freely upwards in summer and autumn to maintain even ripening and reduce a build-up of humidity. Sometimes, however, it is more practical to let the rows follow the contours of the land. In very few sites will all the above objectives be achieved and the grower will have to make some compromises. A vineyard where no such compromises are required is shown in Figure 6.3.

Planting on Slopes

Two methods may be used for establishing vineyards on slopes. Firstly, the rows may run straight up and down the slope. This is generally recommended if the slope is not too steep. Where rainfall can be intense, where soil is not especially stable and the slope is above 7-10°, terracing is often preferred.

It is important to remember that light interception and heat accumulation are promoted

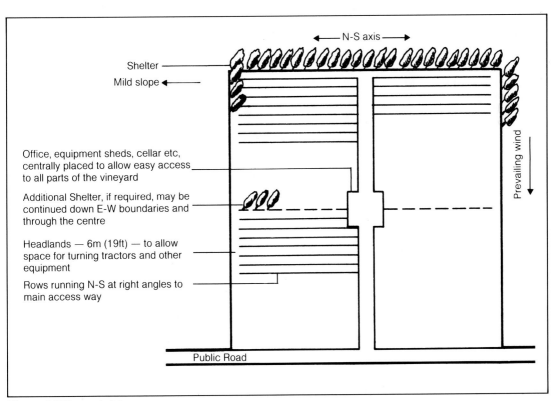

Figure 6.3 Vineyard plot

and frost risk reduced if the slope faces north in the southern hemisphere and south in the northern hemisphere.

Non-terraced vineyards

Planting on mild slopes up and down the incline does not have special needs in terms of preparation and management. Nevertheless if a vineyard has been so planted but erosion commonly occurs after heavy rain, grassing down between the rows can be beneficial.

Terraces

Terraces are considered to be economic when the slope is between 7° and 21°. Soils should be of adequate depth to provide sufficient root growth and the slope must not contain too many natural impediments such as large rocks.

Construction

• The first task on any proposed site is to remove all trees, shrubs, and their roots — plus any other obstacles such as rocks or boulders. The area should then be smoothed by ploughing, discing, harrowing, etc and any gullies or similar features should be filled.

• A system of drainage channels needs to be established which will take water from the high run-off positions so it does not flow through the vineyard. There must be provision for adequate drainage for water-saturated areas and appropriate outlets for water flowing from the block.

• The ground should then be surveyed and a detailed plan of the proposed terraces drawn. Generally, blocks which follow the contours should be no more than 200m long and adequate access ways need to be provided between them.

• Terraces commonly have one or two rows of grapes, examples of which are shown in Figure 6.4 (see also the photograph of the Austrian vineyard in Chapter 1). The distance apart of the two rows, or the distance of the single row from the bank, will vary according to the machinery to be used for cultivation, spraying and trimming. Terraces are prepared using a specifically-adapted bulldozer with a side blade. This should be retractable for travelling, and be capable of maintaining a 45° angle for developing the terrace shoulder.

Planting and Maintenance

• As with any vineyard, a soil analysis prior to planting should be carried out, followed by fertiliser applications to correct major deficiencies.

• Time needs to be allowed for the soil to settle on new terraces prior to planting with vines.

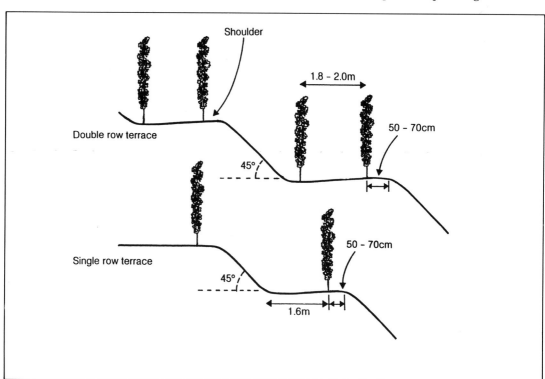

Shoulder

1.8 – 2.0m

Double row terrace

50 – 70cm

45°

Single row terrace

50 – 70cm

45°

1.6m

Figure 6.4 Double and single row terraces for vines

Terrace shoulders should be sown with grass and clovers as soon as possible.
- Vines are planted 50-70cm from the terrace shoulders and staked. A black plastic mulch is often recommended and trickle (drip) irrigation is installed if water deficits are expected.
- The vineyard, depending on growth, is normally trellised in the first or second winter after planting. The trellis should be strengthened at places of greater tension by the deeper placement of posts, greater post density, and by driving the posts at an angle away from the slope.
- Clean cultivation is recommended to provide a coarse surface structure which will facilitate water penetration. If the rainfall is adequate, cover crops may be used annually to maintain soil structure and to limit soil erosion.
- Any grass should be mown then controlled by desiccant sprays, such as paraquat, at the time of greatest water deficit.
- The control of moisture and nutrient levels is especially important on slopes. Regular soil and leaf analyses are recommended and fertiliser is best incorporated to a depth of 30-40cm.
- Terraces, drainage channels, and all access ways should be regularly checked for damage and repaired before serious erosion can occur.

Benefits

Properly constructed terraces should provide a number of benefits, such as:
- The greater air movement found on slopes which reduces frost risk and limits the occurrence of pests and diseases.
- The ability to use modern machinery.
- Adequate control of erosion.
- Under optimum moisture and soil conditions, a terraced vineyard will achieve similar or improved economy due to greater plant density. The greater costs of establishment can be offset by greater yields and lower land value.

The Propagation of Vines

Seeds of grapes are never used for propagation since the plants they produce are different from their parents and usually inferior. Seeds are only used by breeders to produce new varieties, and many thousands of seedlings need to be grown before improved varieties can be selected.

The most common ways of propagating grapes are by using cuttings or grafts. Cuttings constitute the simpler method and are used where *Phylloxera* and/or nematodes are not serious problems.

Cuttings

These are taken in winter from canes produced from the previous year's shoots. Normally canes are at least pencil thickness, but thinner cuttings can be made as long as the wood is healthy and not green. Non lignified (non-woody) green tips of shoots or laterals, formed when the shoots grow late in the season, are unsuitable. The preferred length of cuttings is 25-35cm (10-14in) with three to five buds. Smaller cuttings are only used if material is scarce. Both large and small cuttings may be grown in a nursery, large ones only are suited to planting directly into a vineyard. The nursery should be placed in a sheltered position with an adequate supply of water. Rows can be as close as 60-90 cm (2-3ft) and, in the rows, cuttings may be 10-15cm (3-5 in) apart. If the cuttings are placed in the nursery in spring, and the whole area sprayed with a mixture of paraquat or glyphosate with a pre-emergence herbicide (seek local advice) before bud burst, weed control will be much simpler. Alternatively the cuttings may be placed directly through black polythene as shown in Figure 6.2.

During the subsequent season, especially in spring while the roots are becoming estab-lished, the soil should not be allowed to dry out. Later, if the shoots become too vigorous, they can be trimmed and root wrenching in mid-late summer can encourage roots to develop close to the base. Such plants are easier to handle and are more suited to transplanting. Wrenching can be done with a spade, although there are special devices which can be drawn behind tractors to do the same job.

After wrenching, the soil must be kept moist and, if both trimming and wrenching are to be done, it is an advantage to do both at the same time.

Grafting

Where *Phylloxera* and nematodes are present, grafting to resistant rootstocks is practised. More skill is needed for grafting than the taking of cuttings and it is usually done by specialist nurserymen. However, growers who wish to learn these skills will find detailed instructions in the standard reference-books on general propagation and viticulture. See Notes and References at the end of Part II.

CHAPTER 7
TRAINING GRAPE VINES

Background

In its natural habitat, the forest, the grape vine, like other perennial lianes, such as kiwifruit has developed strategies for survival. Some of these are listed below.

- Drought tolerance which is aided by the vine's strong and vigorous root system.
- Perennial invigoration — this is the ability to store surplus energy produced by leaves in the vine's extensive root, trunk and branch structure. This can be re-used in the next season to promote active growth and increase the overall size of the vine.
- Seasonal invigoration — this occurs when the vine has produced a low crop (common in young, juvenile plants) and diverts its energy into further shoot elongation and root growth.
- Phototrophy — this means the plant's attraction to light. It encourages the vine to divert energy to shoots which are positioned in open spaces.
- Apical dominance and other physiological phenomena which encourage more vigorous shoot growth at higher points in the vine and at shoot tips.

These growth principles are beneficial to the plant in the wild, but can cause problems to growers under monocultural conditions, especially if these growers are unaware of their significance. Growers will cope better with the vine's natural habits if they aim to produce a mature vine where the balance of perennial woody material, seasonal shoot growth, and the annual crops remain as stable as possible. This can be best achieved if accumulation of surplus energy is avoided by two important practices:

1. Encouraging fruitfulness by developing a non-shaded canopy and

2. Removing non-fruitful shoots as soon as possible after bud burst.

Under these conditions the annual photosynthetic production is limited to that needed to:

1. maintain but not expand the vines structure,
2. ripen the evenly-distributed crop fully, and,
3. ensure next season's shoots and crops are similar to the one just completed.

Some of these principles will be discussed again later in the text.

Trellising

There are numerous trellis structures for supporting grape vines, yet the most common in cold climates is the simple vertical type shown in Figure 7.1.

The height is from 1 to 2 m (3-6 ft) and the base wire supports the permanent part of the vines. The plant on the left shows the vine in the winter while, on the right, summer shoots with leaves grow up through the wires above the base.

The factor which, more than any other, can increase yields of a crop is the amount of leaf area exposed to sunlight. The combination of trellis height and the distance apart of the rows will determine the effective leaf area exposed. As rows are moved closer together, the leaf area increases but, ultimately, shading increases to such an extent that the additional leaf area is ineffective or even counter-productive. As a rule, benefits will accrue on flat ground if the distance apart of rows is $1-1^1/2$ times the height of the leaf canopy (not the height of the trellis). Moving closer than this will have no yield benefit and wine quality will

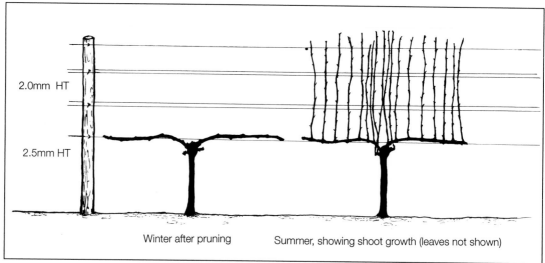

2.0mm HT

2.5mm HT

Winter after pruning   Summer, showing shoot growth (leaves not shown)

Figure 7.1 Vertical shoot-positioned vines

probably decline. Notwithstanding this, growers are often restricted by the machinery available and row distances are usually geared to the width of the widest machine used.

With standard tractors, rows will not normally be less than 2.5 m (8 ft) apart. A trellis should

not be much more than 1.7 m (5 ft 6 in) high, as anything above this height will be costly and difficult for the average vineyard worker to prune and train. With such dimensions and with rows going north-south, reasonable exposure of all leaves will be gained and a good

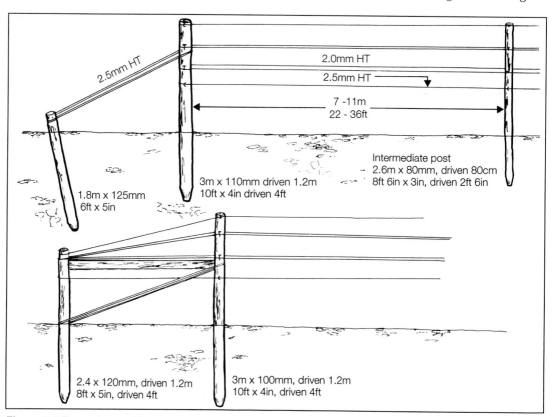

2.5mm HT

2.0mm HT

2.5mm HT

7 -11m
22 - 36ft

Intermediate post
2.6m x 80mm, driven 80cm
8ft 6in x 3in, driven 2ft 6in

3m x 110mm driven 1.2m
10ft x 4in driven 4ft

1.8m x 125mm
6ft x 5in

2.4 x 120mm, driven 1.2m
8ft x 5in, driven 4ft

3m x 100mm, driven 1.2m
10ft x 4in, driven 4ft

Figure 7.2 Two end assemblies

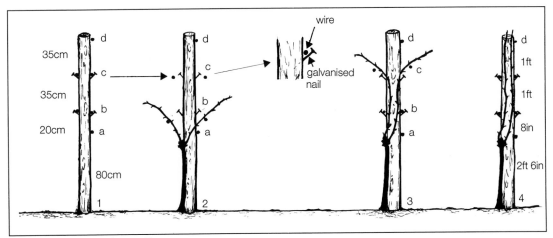

Figure 7.3 End view along trellis

compromise between maximum yield and reasonable efficiency will be possible.

A trellis has two key features: the end assembly and the intermediate posts. The end-assembly enables wires to be strained to the appropriate tension, the intermediates reduce wire sag and provide stability within the row.

A number of end assemblies are utilised, those shown in Figure 7.2 are very strong and generally considered satisfactory for vineyards. A typical trellis design is shown in the same figure. Intermediate posts are close together when the ground is soft and non-supportive, especially in wet weather and/or when heavy crops and vigorous foliage is anticipated. Trellises must be well constructed and unskilled growers must be advised to use contractors. For those wanting more information on trellis construction and design useful material can be found in Smart and Robinson, *1991*.

The Vertical Shoot-Positioned Trellis (VSP)

This is the most common method used in cool climates. It usually consists of a base wire (high-tensile-HT-2.5mm) at 70-90cm and five foliage wires (HT, 2.0mm) above this — see Figures 7.2 and 7.3. If more than one cane is to be laid on either side (cane-pruning, see later), each may be placed on a separate wire. In Figure 7.3 wire 'b' on the opposite side to 'a' would be used. It would be HT, 2.4mm and not movable.

The shoots which grow from the cane on the base wire must be supported by a series of foliage wires. In most cool climates these are a single or double set of parallel wires often arranged as shown in Figure 7.2 and 7.3.

Wires at a and d are fixed. Wires at b and c can

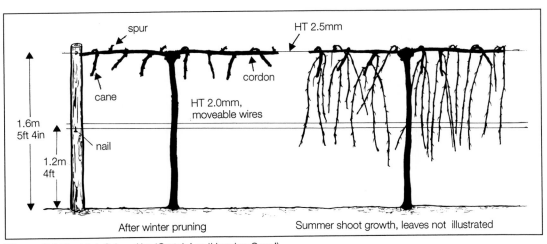

Figure 7.4 The High Sylvoz (the 'Curtain' or 'Hanging Cane')

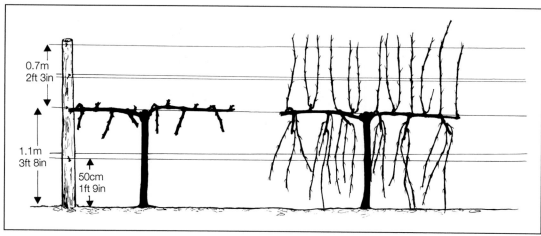

Figure 7.5 The Mid-Height Sylvoz

either be placed over the nails, as at 1, or they can be hung below prior to growth. As the canes grow they fall over the wires b and c, as at 2 and 3, and are then lifted over the nails as shown at 3 and 4.

Other Trellising Systems

The vertical hedgerow system is by no means universal and, before examining methods of pruning, the advantages and disadvantages of some alternatives will be briefly described and compared with the VSP system.

The Sylvoz methods

Methods based on the Sylvoz principle involve training some or all of the shoots produced each year in a downwards direction. These are two versions — the High Sylvoz (also called the 'Single Curtain' or 'Hanging Cane') and the Mid-Height Sylvoz.

The High Sylvoz consists of a high wire and a cordon at 1.6m (5'4"). Canes or spurs on this produce shoots which are allowed to hang downwards. Early encouragement of downwards positioning is advantageous and may be assisted by two movable wires held by downward pointing nails (opposite in direction to 'b' in Figure 7.3).

The Mid-Height Sylvoz uses both upwards and downwards-pointing shoots. It is illustrated in Figure 7.5.

The High Sylvoz is the easiest system to manage, the Mid-Height Sylvoz is more demanding than the VSP but gives better fruit exposure for equal buds per metre of row (see later). The High Sylvoz also gives good fruit exposure due

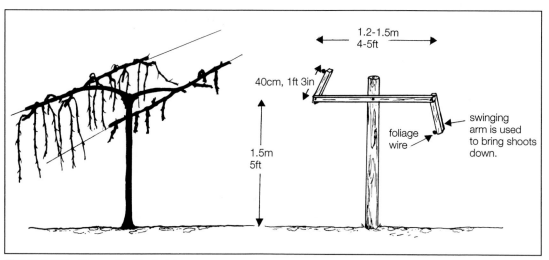

Figure 7.6 The Geneva Double Curtain

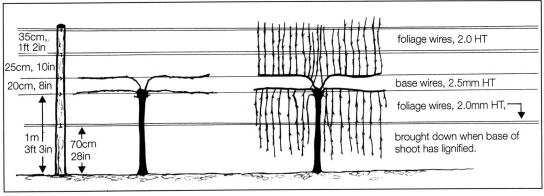

Figure 7.7 The Scott Henry System

to most leaves being below rather than above the berries. Sunburn, especially if shoot positioning is late, may be a problem.

The Geneva Double Curtain (GDC)

This method, developed by Dr Nelson Shaulis of Geneva, New York, is basically a double High Sylvoz. Two cordons are developed in parallel as shown in Figure 7.6.

The system enables a larger leaf area to be produced on a given area and is especially valuable if the equipment used demands wide row-spacing. One major note of caution is that the two hanging curtains must be separate and clearly distinguishable. A poorly-managed, over-grown canopy is a disaster. Providing this does not occur it has the advantages of a single curtain for pruning and fruit exposure.

The Scott Henry

Developed by Scott Henry of Oregon, this is not unlike the mid-height Sylvoz since it utilises both upward- and downward- pointing shoots. It is illustrated in Figure 7.7.

The key to success, as with all methods which utilise downward-pointing shoots, is to position downwards at the correct time — usually just before flowering when the base of the shoot has begun to lignify and is less easily broken by unusual movement.

The Lyre or U Trellis

The Lyre, developed by Dr Carbonneau of Bordeaux as an alternative to the typical Bordeaux close-planting systems, is rather like a reverse GDC.

It is trained like two parallel VSP systems, and, as with the GDC, it is essential to keep the centre open. Some difficulty is experienced summer-trimming the inside which must be done by hand or by specially-adapted machinery.

Row Distances

The Vertical Shoot-Positioned, the Scott Henry and the Sylvoz are usually planted with rows 2.4-2.9m (8-9ft) apart. To accommodate the extra width due to two parallel canopies, GDC and Lyre are commonly 1m (3ft) wider.

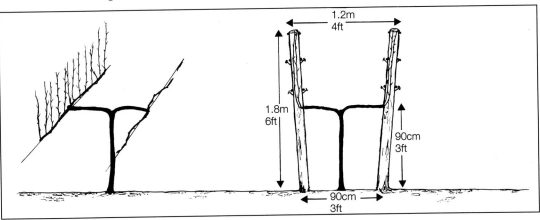

Figure 7.8 The Lyre System

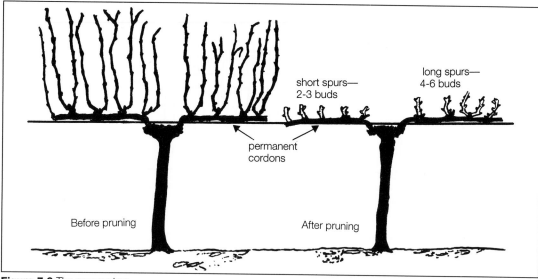

Figure 7.9 The spur system

Pruning

It has already been mentioned that about 90 per cent of the previous-season's growth is removed each winter. It is the one-year-old shoots on which the next season's crop will be produced that are cut off, and it could be suggested, therefore, that 90 per cent of the potential crop is being removed. However, the trellis and the vine could never successfully support this crop level, and removal of shoots is essential. Pruning aims to do the following things:

1. Space the shoots so that each will present its leaves to adequate light.
2. Space the shoots so that air circulation will be encouraged. This will reduce humidity which, in turn, lowers disease incidence.
3. Space the shoots to allow an adequate penetration of sprays for pest and disease control.
4. Provide good replacement shoots for the next winter's pruning.

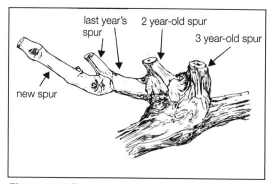

last year's spur   2 year-old spur
3 year-old spur
new spur

Figure 7.10 Spur

5. Select the length and position of the shoot on which the buds have the best potential for fruiting.
6. Achieve an appropriate bud number—per plant or per unit of trellis—to give maximum yield of grapes of optimum quality.

To achieve these purposes it is usual to have a trunk with canes or cordons which are secured to the base wire. There are two ways normally used to achieve this end—the spur system and the cane system.

Spur Pruning

The spur method is simpler than the cane and, in its basic form, it is shown in Figure 7.9.
The detail of each spur is shown in Figure 7. 10.

The advantage of the spur system is that it is simple, and easy to teach to inexperienced workers. The disadvantage is that the basal buds on certain varieties in cool districts produce poor bunches of fruit and the overall crop can be low. Also, after a number of years, some spurs may die and leave bare areas on the permanent arms which may need to be replaced by laying down new canes. It is the method most commonly used by home gardeners and is ideally adapted to growing on fences, pergolas, sides of houses or in glass-houses (commercial or domestic).

While most growers in cool climates prefer the cane system because of its greater productivity, this is not the case in all districts. Growers in new areas should try both the spur and the cane methods before deciding finally on which

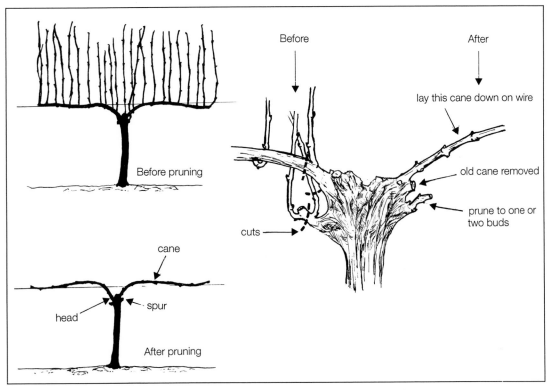

Figure 7.11 The Cane System; 1. Head cane

one to adopt. When making comparisons, the number of buds per metre of row should be identical.

Cane Pruning

Cane pruning is used where basal buds of the canes are not sufficiently fruitful. It is more time-consuming than spur pruning and generally needs a higher level of skill.

The head cane system (Figure 7.11) is most suitable for weaker varieties where the vine can only support a lower number of buds. With

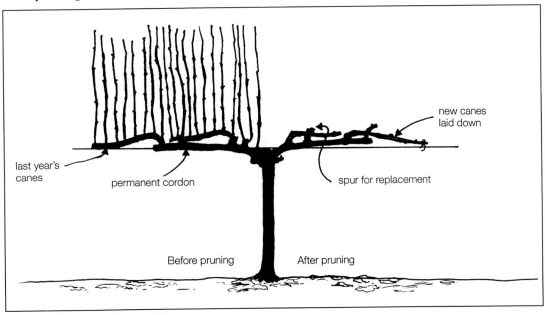

Figure 7.12 The Cane System; 2. Cordon cane

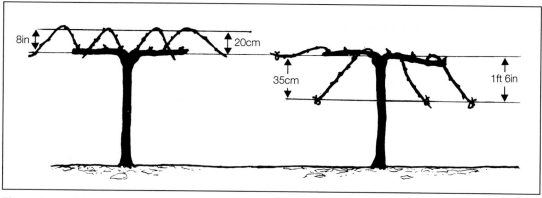

Figure 7.13 Methods used to position canes on wires using cordon cane pruning. The method on the right can be used for Sylvoz training.

more vigorous varieties, such as Cabernet Sauvignon, when higher bud numbers are required, it is more advantageous to use a larger number of shorter canes, rather than two canes of excessive length. Thus in this case the cordon system might be used (Figure 7.12).

The canes are sometimes arched or bent, as shown in Figure 7.13, to use more of the lower part of the trellis, or reduce vegetative vigour while leaving more buds for fruit. Arching and bending can also be used for head canes.

Canes are twisted around the wire and tied securely at the end — see Figure 7.14. The pruner cuts through the last node, removing the bud but retaining the swollen portion— this prevents the tie from slipping off.

Spacing the Vines in Rows

In cool climates, vines are normally spaced 1 to 2 m apart in the rows. Several factors will determine the spacing, although commonly it is the tradition adopted in the area. In general, it may be said that where growth tends to be vigorous, the wider spacing is used. Vigorous growth may be a consequence of the district or the variety, or both. In the alternatives shown in Figure 7.15, three different management methods are used, all finally supplying the same number of buds per metre of row.

Deciding on the Level of Pruning

The yield of grapes will be partly determined by the number of buds left per hectare or acre. The more buds the higher the yield. Some books give recommendations for the number of buds to leave per hectare for specific districts or varieties. Others base the level of pruning on the weight of prunings taken from an average vine in winter. Vines which produce more pruning weight are more capable of coping with a higher number of buds. This method is only reliable when little or no summer trimming is used.

The following recommendations are based on our own experiments and experience, and are appropriate for rows which are approximately 2.6 m (8ft 6in) apart. Bud numbers are expressed as buds per metre (or foot) of row— which overcomes the problem of comparing buds per vine if the vine spacing in the row varies. Bud numbers can be increased by leaving more canes or longer spurs. When high bud numbers are required, the spacing of canes by the methods shown in Figure 7.13 is suggested to avoid congestion and excessive shading of fruit and lower leaves. Divided canopies like the Scott-Henry will achieve similar effects.

Generally, higher bud numbers are used when the variety is normally a poor cropper: for example, Gewürztraminer may only produce 70-100

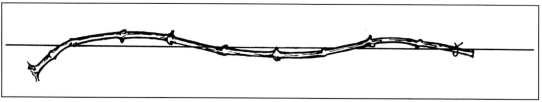

Figure 7.14 Tying a cane on a wire

Figure 7.15 Cane pruning variations for vines of different vigour and spacing—the cane and spur of the one-way variation on the left are alternated in position each year.

grams of fruit for every bud retained after winter pruning, Müller Thurgau will produce up to 400 grams. The following table can be used as a guide to pruning levels.

Such data are appropriate for varieties growing in cool climates in their normal LTI zone. If a low-yielding grape is grown in a warm area then yield will often be higher. For example, Gewürztraminer cropped in a warm situation can be much more productive than the same variety in a cool district.

There are therefore some occasions where such bud numbers can be modified.

The bud numbers shown in Table 7.1 apply when:

1. The grape is grown in its normal LTI zone.
2. The soil is well drained and of low to medium fertility.
3. The vines are well established and have cropped for five years or more.
4. The area has adequate, but not excessive rainfall.

Bud numbers will be increased when:

1. Young vigorous vines are grown on fertile soil and display excessive vigour.
2. Poor clones of low fruitfulness have been planted.
3. Water (natural or by irrigation) is always in plentiful supply.
4. Row spacings are wide and ample space is available to lay down buds.
5. Grapes are grown in warmer situations and the consequent delay in ripening due to higher yields causes no problems.
6. Wine quality is of secondary importance to yield—e.g., for bulk wine.

Bud numbers should be decreased when:

1. Insufficient ripeness and delayed harvest occurs in normal years.

Table 7.1 Recommended bud numbers per metre (foot) for different varieties.

Variety	Buds per metre	Buds per foot
•*High yielding*, e.g. Müller Thurgau Kerner Chenin blanc	15-20	4-6
•*Moderate to high yielding*, e.g. Pinot blanc Sylvaner Riesling Chasselas Pinot gris	20-25	6-8
•*Moderate yielding*, e g Malbec Sémillon Pinot Meunier	25 30	8-9
•*Moderate to low yielding*, e.g. Merlot Cabernet Sauvignon Pinot noir Sauvignon blanc	30-35	9 -11
•*Low yielding*, e g Chardonnay Gewürztraminer	35- 40	11-12

2. Poor cane vigour and inadequate lignification occurs.
3. The wine produced displays poor intensity of flavour and bouquet, and has little varietal character, plus a low sugar-free extract.

Mechanical Pruning

A number of growers in several districts are now using mechanical methods of pruning. Vines so treated are pruned to a spur system and the cut above the cordon is made by some type of mechanical cutting device. The height above the cordon determines the buds left per metre of row. Obviously, the operation cannot be so finely controlled as by hand, but experimentally and practically, it has a number of advocates. The time saved is considerable and yields and quality do not necessarily suffer.

Over time the cordon becomes cluttered with old stubs of canes and growers will, every so often, intersperse a hand-pruning operation.

Viticulturists contemplating mechanical pruning are advised to test it on a portion of the vineyard for two to three years before adopting it generally.

Another mechanical approach is being used successfully in Coonawarra, and Padthaway, Australia. It is called 'minimum pruning' and begins after a normal summer of growth when the excessive removal of canes by cane or spur pruning is not undertaken. Instead the approach used for the summer trimming of shoots is repeated, but the cutters are brought closer to the wires. Next summer the shoots grow outwards and are not trained between wires. The higher number of shoots inhibits excessive growth, and bunch number and berry size are reduced. Grapes are distributed on the canopy perimeter rather than in the shaded interior. Yields tend to be higher than for hand pruning, but, while high cropping levels can sometimes be associated with lower quality, these can be offset by other factors improving quality viz. good fruit exposure and a much increased ratio of skins to juice due to smaller berries (giving better colour in red grapes).

Small-berried, vigorous varieties like Cabernet Sauvignon, Gewürztraminer and Sauvignon blanc are said to have benefitted most from this technique. Economic savings are considerable.

Summer Pruning

Once the shoots pass the highest wire they are usually topped and, if laterals become too dense, they can also be trimmed back. Both these cuts may be achieved by mechanical cutting devices.

If the tops are trimmed just before the flowers are ready to be pollinated (cap fall), fruit set can be improved, but on susceptible varieties second set can be increased. The frequency of summer pruning will depend on the vigour of the vines; where vigour is low, once or twice will be sufficient. Under more vigorous conditions it may be necessary to prune three or four times. Excessive growth may occur in warm humid areas, or where too few buds have been retained. It will be remembered that normally, in

cool areas, the growth of shoots declines when the berries are in their final stage so that the last pruning will normally take place at around *véraison*.

Pruning and Training for Quality

Growers and winemakers in cool climates have a special interest in quality, after all if bulk wine were the aim, the grower would probably be better served by relocation in a warmer climate. Thus the grape grower must be aware at all times of the relationship between his/her practice and the possible quality of the wine to be made from the grapes. Here are three principles which if followed will encourage the attainment of quality.

• Avoid overcropping since it will delay maturity, reducing sugar levels and increasing acidity. The solution is to reduce bud numbers per metre of row — see 'Deciding on the level of pruning'.

• Reduce congestion which will lower quality due to an increased possibility of disease and the effects of shade on juice composition. In high-shade, grape colour is reduced, sugars are low, acids and pH are high, and, in addition, undesirable flavours may be increased.

• Control pests and diseases which are almost always bad for quality.

In the 1980s considerable research was done to elucidate the factors involved in quality and its relationship to light interception. The general conclusion was that measures adopted to reduce shade could pay handsomely in terms of quality enhancement. Many New World vineyards were too vigorous and the training systems being used did not adequately cope with the vigour. The solutions proposed generally entailed dividing the canopy, as in the Scott Henry, Lyre, mid-Height Sylvoz or the GDC. Some examples may illustrate this conclusion.

In the vertically shoot-positioned canopy it has been seen (see Robinson and Smart, *1991*) that a shoot density of more than 15 buds per metre will lead to excessive shade in the canopy. Yet Table 7.1 shows that to achieve adequate and therefore economic yields most varieties must have more than 15 buds per metre.

Cabernet Sauvignon in a VSP at 30 buds per metre will produce a crowded canopy. Divided into two, as in the Scott Henry for example, there are effectively only 15 buds (and shoots) per metre and crowding is reduced to an acceptable level. If, however, in the chosen climate Gewürztraminer were grown and this needed, say, 40 buds to achieve an economic yield, even dividing the canopy may still result in a shaded canopy — especially with this variety which not uncommonly produces more than one shoot per bud. A useful adjunct can then be adopted — this is the removal of leaves from around the fruit zone at or near véraison. This can be done by hand or by machine. Leaf plucking is done routinely by growers who regularly have to deal with over-dense canopies. It is not required in a good open canopy.

Appropriate trimming and topping of vines are essential to achieve a good canopy, with good light distribution.

Close Planting

Many classical European vineyards have spacings much closer than are found in New World areas. In Burgundy, Champagne, Bordeaux for example spacings of 1m x 1m are common. By using such spacings the French did effectively what the New World Viticulturists are now achieving by canopy division. A simple example will explain this: A vineyard planted in the Scott Henry system with canopy height of 1.8m and between-row spacing of 2.8m achieving 15 shoots per metre in each divided canopy will have no better fruit exposure than a VSP planted at a row spacing of 1.4m and height of 0.9m, also at 15 shoots per metre.

Close planting therefore makes obvious sense. However, some essential factors should be considered before utilising close spacing:

- Close spacing demands specialised machinery and since such equipment is not standard in many districts its procurement and servicing may be very costly.
- Close spacing is *not* recommended in fertile soils. Individual plants in close spacing *are* less vigorous but not sufficiently devigorated to compensate for the vigour induced by high fertility,
- The best utilisation of the soil in low-vigour vineyards is in planting on the square. That is why many French growers prefer, say, 1m x 1m to 2m x 0.5m.
- Being closer to the ground vines may gain more reflected heat and maturity may be advanced. Proximity to the soil may exacerbate frost damage in spring and autumn.

Pruning and training is essentially the same as described above, with some minor variations which are illustrated in Figure 7.15 (left).

Controlling Vigour

It will be now recognised that excess vigour in vines is not a desirable feature. Several management practices can be used to reduce vigour, some have already been introduced and others will be described later. In brief these are:

- choosing an appropriate *terroir* and climate
- choosing less-vigorous rootstocks
- planting grass between rows to compete for nutrients and water
- training larger rather than smaller vines
- distributing shoots over a wide area (canopy division)
- irrigating wisely, using fertilisers sparingly, if at all
- shoot thinning
- chemical use
- girdling.

The latter three options will now be discussed.

Removal of non-bearing shoots. The length of a shoot has an inverse relationship to the weight of the crop carried on the shoot: Effectively, once the bunch of grapes rapidly expands, nutrients are diverted from the apex to the cluster and shoot growth slows down or stops. Shoots with no clusters grow the longest, they cause congestion and because there are no grapes to absorb the photosynthates produced by the leaves, the excess is diverted to the rest of the plant including the roots. This excess promotes further vigour and compounds the problem. The solution is for the grower to remove shoots with no crop or very small crop before they are 20-30cm long. Vigorous vines so treated can often become easily controlled and yield and quality can both be improved.

Growth-regulating chemicals. A number of chemicals will reduce vigour some, such as *chlormequat (CCC)*, effective on vinifera grapes, and *daminozide* on labruscas, are now generally prohibited. *Paclobutrazol* will reduce shoot growth but its value for vines has still to be satisfactorily demonstrated.

One compound *ethephon* ('Ethrel') is effective and probably benign. Ethephon (300ppm a.i.) sprayed on vines stops growth of the shoot tips and laterals, the effects last for 4-6 weeks. Cropping may be reduced if applied before capfall. A *tentative* recommendation is: apply first at 3 weeks after capfall — direct spray to shoot tips and avoid fruit zone. Four-six weeks later spray the whole plant. Further sprays are generally not required and even the second is not always necessary. Ethephon sprayed after veraison will advance maturity — a fact which will be mentioned later. Check in your district if any restraints are placed on ethephon use in grapes.

Girdling. Girdling is the removal of a ring of bark around the trunk or the canes of vines. Applied when shoots are 10-20cm long, girdling will reduce shoot growth for several weeks, applied 1-2 weeks before capfall, it may promote fruit set. Special tools are available to assist the operation.

Girdling is seldom used but deserves to be considered more seriously. The problem is that the response varies, but it generally only works o varieties which in the district set poorly a wide girdle in a cool climate can often severely debilitate the vine. It would pay growers to experiment on a few vines using narrow girdles (3mm) in the first trials. Such trials may pay handsomely.

Training the Young Vines

In this section the development of the vine is followed from the time it is planted to when it is at a mature stage and will be trained by one of the previously-described methods.

The cutting, rooted or grafted plant, is inserted into the soil to about two-thirds of its length and is sometimes lightly covered with soil to protect it from spring frosts. The soil is removed three to four weeks after growth has commenced.

Only one bud is allowed to grow and this is tied to a stake, or trained up string tied to the wire (Figure 7.17).

The growth of young vines depends on water availability, soil moisture and fertility plus climate. These must be optimised to encourage early vigour. The alternative methods of training weak, moderate or vigorous vines are shown in Figure 7.16.

Sometimes, in the first year, the shoot will not reach the wire, in which case it is normally pruned back the next winter to three or four buds. If growth is very vigorous and reaches the wire early in the season it may be either trained one way along the wire or cut off below the wire to encourage the shoot to divide. If growth is moderately vigorous and reaches the wire later in the season it will be cut off in winter and tied to the wire. Subsequent treatments of weak, moderate or vigorous vines are shown in the lower part of Figure 7.16. It will be seen that vines with moderate to vigorous growth have produced a small crop in the second season after planting. After the second winter—the third growing season—these fruiting vines are nearing full production. Weak vines will need a further season to reach this stage.

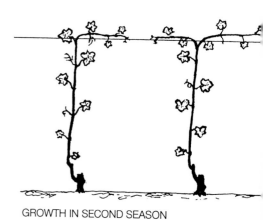

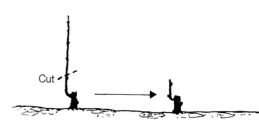

WEAK GROWTH

Cut

PRUNING IN THE FIRST WINTER AFTER PLANTING

GROWTH IN SECOND SEASON

5-6 buds   3-4 buds

PRUNING IN THE SECOND WINTER AFTER PLANT

Figure 7.17 Early training of vines with different vigour

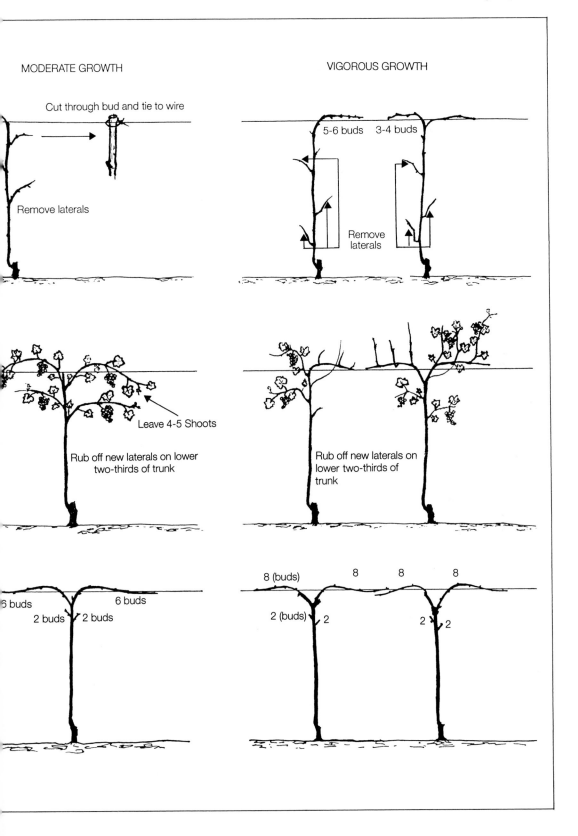

MODERATE GROWTH

Cut through bud and tie to wire

Remove laterals

Leave 4-5 Shoots

Rub off new laterals on lower
two-thirds of trunk

6 buds   6 buds
2 buds ⅄ 2 buds

VIGOROUS GROWTH

5-6 buds   3-4 buds

Remove
laterals

Rub off new laterals on
lower two-thirds of
trunk

8 (buds)   8   8   8
2 (buds) ⅄ 2   2 ⅄ 2

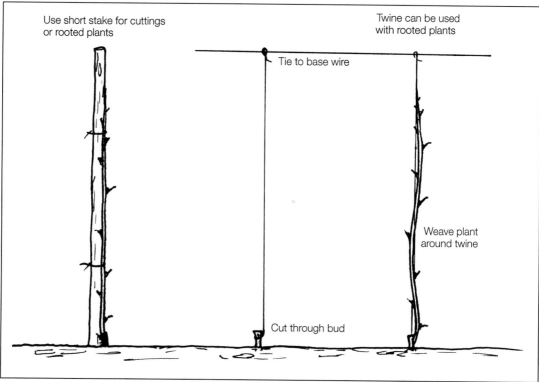

Figure 7.17 Supporting the shoot

Details of Figure 7.16 are illustrated in Figures 7.18, and 7.19 . The vine (top, in Figure 7.18) has been pruned back to the base wire, that to the right, a little more vigorous, has been trained a little way along the vine. All buds below the top five (left) or the top nine (right) should be removed at this stage.

The vine in Figure 7.19 was trained to the wire the winter before last. This year six to eight buds on each arm have been laid down together with two 2-bud spurs.

Figure 7.18 The second season after planting

Figure 7.19 The third spring after planting

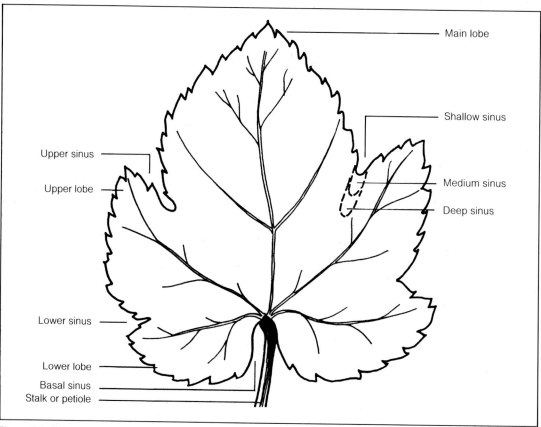

Figure 8.1 Main parts of the vine leaf

CHAPTER 8

GRAPE VARIETIES FOR COOL CLIMATES

Careful selection of climatic zones, sites and soils will be of little value to the viticulturist or winemaker if unsuitable varieties are planted.

In this chapter we list all major and many minor cool-climate grape varieties. Where alternative names are common, these are shown in brackets.

Descriptions of grapes are supported by notes on the desirable aspects of husbandry—data which will assist a grower to choose varieties which would be suitable for a specific environment. Details are summarised in a table at the end of the chapter.

It should be remembered that some variations in the characteristics described can occur due to factors such as climate, soils, clones, pests and diseases and virus infection. The comments and descriptions relate to healthy mature plants, trained and grown in a satisfactory manner under good vineyard conditions. Young plants also vary in minor ways from mature plants.

As parts of the leaf are sometimes mentioned, Figure 8.1 is provided to serve as a means of identification. The undersides of leaves often have a woolly type of growth which is referred to as pubescence.

Riesling grapes close to harvest

Descriptions and terms used

The Vine		
	Vigour	Will be described as very vigorous, vigorous, moderate vigour, little vigour.
	Nature of shoots	Mainly refers to the colour of young shoots and mature canes. Where significant, internode length is mentioned.
	Pruning method	The comments here are for the varieties of vines that are caned pruned. In this context a short cane has 6-8 buds (nodes) and a long one 10-12. Where spur pruning has been found to be satisfactory growers will disregard these comments.
	Rootstocks	Any known preferences will be mentioned.

Cabernet Sauvignon

A premium black grape trom Bordeaux, where it is grown together with other grapes for the production of the famous Médoc and Graves clarets. In cool climates this variety ripens late in the season, giving small to medium yields of high-quality grapes. However, this grape will not ripen in very cool districts, for example: northern France, Germany, England, and the cooler parts of the USA. Fullness of varietal character, depth in bouquet and flavour, coupled with high acidity and tannin content are the main characteristics of Cabernet-based wine. The quality of the wine compensates for the lower yields associated with this vine. The Cabernet leaf is very distinctive: dark green and rough on the surface with very deep sinuses; reddish shoot growth in spring is typical.

Agronomy		
	Soils	Suits most soils
	Vigour	Vigorous
	Pruning	Medium to long cane pruning preferred with a high proportion of old wood to young canes (i.e. consider cordon-cane pruning)
	Disease susceptibility	Downy mildew—fair
		Powdery mildew—high
		Bunch rot and *Botrytis*—low
	Wet weather	Good resistance to wet weather at harvest
	Rootstocks	To control vigour and improve set choose those with low to moderate vigour.
	Yields	Low, but can be increased with good pruning and selection of improved clones.

Chardonnay (Pinot Chardonnay)

A premium-quality white grape variety of Burgundy, Champagne and other districts of central Europe. It is now becoming popular in areas outside Europe. There has been some confusion between this variety and Pinot blanc. Chardonnay can be recognised by its distinctive, wide open, basal sinus and by its yellow-green grape. The wine has a fine and very distinctive varietal aroma and flavour, although in soils lacking calcium it is less aromatic. Generally not suited to heavy clay loams. Poor clones are found in a number of areas and it is important that good selections are chosen.

Agronomy		
	Soils	Likes well-drained, fertile but drier calcareous soils. Heavy nitrogen fertilisation not recommended since set becomes poor and grapes are more susceptible to rots
	Vigour	Moderate
	Pruning	Medium to long cane pruning
	Disease susceptibility	Downy mildew—low
		Powdery mildew—moderate
		Bunch rot and *Botrytis*—moderate to high
	Wet weather	Susceptible to rain at harvest
	Rootstocks	Vigorous ones preferred, except in fertile soils
	Yields	Low to moderate—depending on clone and climate

Chasselas (Chasselas Doré or Golden Chasselas)

This white grape is grown in the Alsace and Loire regions of France and other cool regions such as Switzerland, Austria and Germany. It is sometimes found under local names such as Fendant, Gutedel, Sweetwater and others. The large grapes of early-maturing Chasselas give high yields of ordinary wine with slight aroma, neutral flavour, low acidity, a low alcohol level, yet with pleasant character, but without much distinction The variety is also suitable as a table grape and for grape juice, but perhaps its major- value is as a low-acid wine for blending in cool-climate regions.

Agronomy		
Soils	Yields well on a number of soils, although most suited to lighter and deep soils	
Vigour	Low to moderate with thin shoots	
Pruning	Short to medium canes	
Disease susceptibility	Downy mildew—moderate to high	
	Powdery mildew—low to moderate	
	Bunch rot and *Botrytis*—moderate	
Wet weather	Fair resistance	
Rootstocks	Moderate vigour preferred	
Yields	Moderate to high	

Chenin blanc

A premium white wine variety of the middle Loire in France; best known in the areas of Anjou and Touraine as Pineau de la Loire or Blanc d'Anjou. Moderate to high yields, and a quality wine in the dry, sweet or sparkling style, make Chenin blanc one of the most popular French varieties in California and more recently in South Africa, where large plantings of this grape are encouraged. In the Loire, Chenin blanc is used for numerous styles, giving premium wines from late-harvest grapes, to produce the sweet wines of Vouvray and Saumur. The wine is light and retains a pronounced varietal character with medium acidity. Late-harvest grapes give distinctive wines with complex character.

Agronomy		
	Soils	Calcareous soils are, in France, said to improve the quality of Chenin blanc wine considerably, although this has not been confirmed in South Africa or California where it appears less demanding in relation to soil
	Vigour	Vigorous
	Pruning	Medium to long canes with a high proportion of old wood to new canes (consider cordon-cane pruning)
	Disease susceptibility	Downy mildew—medium to high Powdery mildew—moderate to high Bunch rot and *Botrytis*—moderate to high
	Wet weather	Susceptible
	Rootstocks	Use those with moderate vigour
	Yields	Moderate to high

Gewürztraminer (Savagnin Rosé)

A clonal selection of the Traminer variety of Northern Italy, Alsace and Germany. Gewürz means spicy and the variety gives both white and pink berries, often in the same bunch. Gewürztraminer matures early, giving low to moderate yields of distinctively-spicy wine, pronounced varietal aroma and flavour, and low acidity. The Traminer leaf is very distinctive, being small, round, dark green and rough with shallow lobes.

Agronomy		
	Soils	Likes deep fertile soils, use nitrogen with caution since excess causes poor set. Has a tendency to exhibit chlorosis in calcareous soils
	Vigour	Moderate— has short internodes, lacks upright growth. There is a tendency for multiple shoots to be produced from the buds and some summer thinning of the shoots is advisable
	Pruning	Use long canes with a higher proportion of older wood in a cordon-cane system
	Disease susceptibility	Downy mildew—fair Powdery mildew—high Bunch rot and *Botrytis*—high
	Wet weather	Susceptible to wet weather at flowering and ripening, causing, respectively, poor set and splitting of berries
	Rootstocks	Low to moderate vigour preferred
	Yields	Low

Kerner

This is possibly the most outstanding German crossing made in recent years. It was bred and selected at the Weinsberg Institute in 1929 and received a patent in 1969. It quickly became very popular so that by 1992 7826 ha (19,330 acres) were planted in Germany. Its principal advantages are: suitability to cool climates, not being very demanding of soils, having fair resistance to autumn frosts and producing fine quality wines. Its disadvantages are its susceptibility to Powdery Mildew and *Botrytis* and the presence of an unknown condition 'The Kerner disease' which is causing problems in its native country.*It ripens mid-season, the wine has a distinctive peach-like bouquet, and fine flavour with mild acidity retained at higher levels of ripeness. Kerner is well suited to the late-harvest Spätlese, Auslese, etc styles and grapes reach a suitable stage just prior to Riesling. It develops well with bottle maturation .

---

* Kerner disease. This refers to dieback of mature vines found in a number of German plantings. It is thought to be exacerbated by soil compaction and excessive soil moisture. Growers should be careful not to plant material gained from infected areas.

Agronomy		
	Soils	Not demanding, but avoid excessively heavy, moisture-retaining soil types
	Vigour	Moderate, tends to have high number of lateral shoots which increase shoot density
	Pruning	Use long canes
	Disease susceptibility	Downy mildew—low
		Powdery mildew—high
		Bunch rot and *Botrytis*—high
	Wet weather	Poor resistance at ripeness, since berries detach easily from bunch stalks
	Rootstocks	Use more vigorous ones like 5BB in average soils, in fertile soils use SO4]
	Yields	Moderate to high and regular, 20 per cent above Riesling

Malbec

One of the principal grape varieties used for claret wine production in the Bordeaux district of France where it is known as Cot. This variety is often said to yield irregularly, yet in some regions it does well (Bordeaux and Argentina), due possibly to the use of better clones and rootstocks. It produces a wine of soft, full character, well-coloured and with considerable amounts of tannin. It is a useful variety for blending with Cabernet grapes. Its cultivation tends to be restricted to areas of low humidity since it is susceptible to wet weather at flowering and maturity.

Agronomy		
	Soils	Well-drained fertile soils are best
	Vigour	Medium
	Pruning	Use medium to long canes
	Disease susceptibility	Downy mildew—fair to high
		Powdery mildew—fair
		Bunch rot and Botrytis—high
	Wet weather	Poor set and berry splitting are the consequences of wet weather at flowering and ripening
	Rootstocks	Low to moderate vigour
	Yields	Moderate to high

Merlot

Another major black grape of Bordeaux used in the production of claret. In Médoc it is planted to a lesser extent than Cabernet, in Pomerol and St. Emilion it is predominant. Merlot is used to improve colour, flavour, texture and balance in the wine, and gives considerable depth and softness to all Cabernet-based wines; it retains fair acidity and tannin. Merlot has fair wet weather resistance and has proved useful in warmer sites.

Agronomy		
	Soils	Does well on a number of soil types although deeper warmer soils are considered best
	Vigour	Moderate
	Pruning	Use medium to long canes
	Disease susceptibility	Downy mildew—fair
		Powdery mildew—fair
		Bunch rot and *Botrytis*—fair to low
	Wet weather	Fair to good resistance
	Rootstocks	Low to moderate vigour
	Yields	Moderate—but may be low in cool climates, especially in plants derived from cuttings (i.e. not grafted)

Meunier (Pinot Meunier, Müllerrebe, Schwarzriesling)

A quality black grape of central Europe, grown in Champagne and certain German areas. It is a rather similar variety to Pinot noir, though buds burst slightly later and its smaller grapes ripen with a lesser amount of acidity. The Meunier wine has a soft, aromatic character, with less depth of colour than Pinot noir. It is perhaps most suitable for blending with similar but more acid wines of Pinot stock, such as Pinotage or Pinot noir. Young shoots are distinctively white tipped and mature canes are greyish brown.

Agronomy		
	Soils	Well-drained soils preferred, suits calcareous soils
	Vigour	Moderate
	Pruning	Medium to long canes, avoid dense foliage by summer pruning
	Disease susceptibility	Downy mildew—low to fair
		Powdery mildew—fair to high
		Bunch rot and *Botrytis*—fair to high
	Wet weather	Induces disease at harvest time
	Rootstocks	Moderate to vigorous
	Yields	Moderate

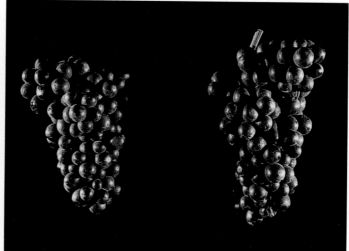

Müller-Thurgau

An early crossing released in Germany in 1924. Initially very popular in Switzerland, it is now a major variety in Germany, Austria, the United Kingdom and other cool-climate countries of Europe, and also in New Zealand. The wine, although not as fine as Riesling, has a distinctive character and a fine balance of acidity, flavour and aroma. This grape is best suited to positions where Riesling ripens in only the best vintage years. Its high yields and early ripening have contributed to its popularity. Some newer Riesling and Sylvaner crosses are, to some extent, replacing it in the latest German plantings. The grape is sometimes called Riesling Sylvaner.

Agronomy		
	Soils	Does well on fertile soils but is not suited to calcareous soils—which tend to induce chlorosis—or to very dry soils
	Vigour	Low to moderate; in dry regions insufficient replacement canes can occur in older vines (10 to 15 years), leading to reduced yields of small berries in loose bunches which ripen late. In the UK yields are much lower and vigour is greater than in heavy-cropping districts.
	Pruning	Use short to medium canes
	Disease susceptibility	Downy mildew—high
		Powdery mildew—high
		Bunch rot and *Botrytis*—fair to high
	Rootstocks	Likes more vigorous rootstocks, however Kober 5BB is not suitable since it predisposes the plant to potassium deficiency. Kober 125A is more suitable
	Wet weather	Susceptible
	Yields	High (Low in UK)

Muscat Ottonel

A quality white grape variety grown in Alsace and elsewhere in central Europe. Consistent cropping, low acidity and early ripening, make this a popular variety. Muscat Ottonel wine has fine aroma and a pronounced varietal flavour and is often used to provide a Muscat aroma to blended white wines. Although it requires careful selection for position, soil, and cultivation, it could prove suitable for many areas. Short internodes are typical.

Agronomy		
Soils	Deep fertile soils are best, calcareous soils induce chlorosis. Drought conditions produce insufficient growth
Vigour	Low to moderate with a tendency to spread rather than grow upright: thus smaller systems such as the head cane are recommended
Pruning	Use short canes
Disease susceptibility	All diseases—fair
Wet weather	Moderately susceptible
Rootstocks	Moderate to vigorous. Choose those like 26 G or SO-4 which improve set: poor set is an inherent problem with this variety
Yields	Moderate

Pinot blanc

A premium French white grape variety of southern Burgundy and Alsace. It is cultivated elsewhere in central Europe, sometimes under the names Klevner or Weissburgunder and gives moderate yields of early-ripening grapes. The wine retains fair acidity with fine varietal flavour and aroma.

Agronomy		
	Soils	Likes well-drained stony or calcareous soils. It has moderate resistance to drought
	Vigour	Moderate
	Pruning	Use long canes
	Disease susceptibility	Downy mildew—low
		Powdery mildew—fair to high
		Bunch rot and *Botrytis*—fair to high
	Wet weather	Fair resistance
	Rootstocks	Does well on vigorous rootstocks
	Yields	Moderate to high

Pinot gris

An early-ripening, quality white grape grown in all central Europe, often under such names as Rülander, Grey Burgunder and others. Notably successful in Rheinpfalz, Alsace and the Loire. It is thought to be a bud mutation of Pinot noir, and it sometimes produces dark-blue grapes, rather than the normal greyish-blue berry. Yields reasonably well, especially new selections such as Klosteneuburg and Hauser H-1 . This grape retains low to fair acidity and a high sugar level: in better positions it gives fine varietal aroma and flavour. Because of high extract levels it often provides a 'Late Harvest' character to wines.

Agronomy		
	Soils	Grows on most types of well-drained soils. Has good tolerance of drought conditions
	Vigour	Moderate
	Pruning	Use long canes
	Disease susceptibility	Downy mildew—low
		Powdery mildew—low
		Bunch rot and *Botrytis*—low
	Wet weather	Fair resistance
	Rootstocks	Use vigorous rootstocks
	Yields	Moderate, sometimes good, especially with new clones

Pinot noir

A premium quality black grape from the Burgundy and Champagne regions of France which is also popular in other cool areas of central Europe. Known sometimes as Blauer Spätburgunder or Cortalloid, it also has other local names. A large number of clones of this variety are grown; one of these, Davis Clone 19, was incorrectly called Gamay de Beaujolais inCalifornia. The better selections, when grown in suitable conditions, give moderate yields of fine wine with considerable softness, depth and elegance. Colour is developed early, well before it is ripe, but nevertheless the wine often has insufficient colour in warmer climates; it retains fair acidity.

Agronomy		
	Soils	Likes well-drained deeper soils
	Vigour	Moderate
	Pruning	Long canes — retain open canopies by summer pruning to induce colour development
	Disease susceptibility	Downy mildew—fair
		Powdery mildew—fair
		Bunch rot and *Botrytis*— fair to high (rots quickly if damaged by insects or birds)
	Wet weather	Fair resistance
	Rootstocks	Moderate to vigorous
	Yields	Low to moderate

Riesling, (White Riesling, Rhine Riesling, Johannisberg Riesling or Petit Riesling)

A premium white grape variety of the Mosel and Rhine regions of Germany, also grown in Alsace and other premium wine districts of central Europe. In Australia, California and New Zealand it has successfully produced fine wines of outstanding quality and breed: having a noble aroma and depth of flavour, yet retaining acidity. In cool areas it ripens late, and in cold seasons, or on poor sites, it often has low sugar and high acid levels. Yet, in the same areas, after a good summer and autumn, it produces a wine that is unmatched in quality. The very tight bunch with spotted berries and red-tipped spring shoots are distinctive.

Agronomy

Soils The quality and quantity of grapes will depend on the suitability of the variety to the soil type. In excessively fertile soils it has a tendency to poor set and gives wine of little character. Lighter, well-drained, warm soils of stony alluvial or volcanic origin with good nutrient balance, especially with respect to potassium and magnesium, produce the best wine

Vigour Moderate

Pruning Use long canes

Disease susceptibility Downy mildew—fair

Powdery mildew—fair

Bunch rot and *Botrytis*—high

Wet weather Poor resistance at flowering and ripening; susceptible also to sunburn

Rootstocks Rootstocks are selected according to soil type; SO4 for stony soils, 5-C for calcareous loams and G-26 for deeper volcanic soils. Kober 5BB is never used for Riesling, since it leads to poor set and wine quality

Yields Generally considered to be a low cropper but new clones have given moderate to high yields. The best-known clones in Germany are Gm 94, Gm 110, Gm 119, Gm 239, and Hochzuchtriesling

Sauvignon (Sauvignon blanc)

A premium white grape of Graves and Sauternes in Bordeaux, also grown in the Loire and elsewhere in central Europe. Known sometimes as Petit Sauvignon or Muscat-Sylvaner, although other names are also used. The yield and quality of Sauvignon varies, depending largely on the area and training of the vines. Younger vines, especially if trained on a low trellis, give low yields; however, with a high trellis, the more vigorous growth of older vines considerably improves yield and often quality. Sauvignon is capable of producing wines of fine quality (dry or sweet) with strong varietal character and with considerable depth and complexity. The grape ripens late with fair acidity, sometimes with quite high sugar content, fine aroma and flavour. Generally, small and tight bunches are produced on short stalks. The most impressive examples of Sauvignon in the New World are the wines of New Zealand.

Agronomy		
	Soils	Prefers gravelly or sandy loams and does not do well in very dry or calcareous soils. Deep fertile soils produce excessive growth and poor crops, and make the berries prone to rot
	Vigour	Vigorous to very vigorous
	Pruning	Medium to long cane pruning is used with a high proportion of older wood—use cordon cane rather than head cane—summer pruning is important to assist air circulation
	Disease susceptibility	Downy mildew—fair Powdery mildew—fair to high Bunch rot and *Botrytis* — high. Late *Botrytis* infection in cool dry conditions (Noble Rot) can produce sweet, powerful and distinctive wines: Sauternes being a good example
	Wet weather	Poor resistance
	Rootstocks	Low to moderate vigour preferred
	Yields	Low to moderate, depending on training

Sémillon

The Sémillon white grape is grown extensively in the Graves and Sauternes regions of Bordeaux, producing both fine and dry or late-harvested sweet wines of considerable finesse and aroma. This grape is also grown in Australia where large plantings are reported and where it is known sometimes as Hunter Riesling and White Madeira; in South Africa it is known as 'Green Grape'. The variety yields moderate crops with larger grapes than Sauvignon (they are grown together in Bordeaux), and ripens slightly later. The Sémillon wine has a soft character, fair acidity in cool climates, and a distinctive herbaceous aroma and flavour.

Agronomy	*Soils*	Prefers lighter soils
	Vigour	Moderate to vigorous
	Pruning	Use long canes and summer prune in more humid areas
	Disease susceptibility	Downy mildew—fair
		Powdery mildew—fair to high
		Bunch rot and *Botrytis*—high, but see comments under Sauvignon blanc
	Wet weather	Poor resistance, also susceptible to sunburn
	Rootstocks	Moderate vigour
	Yields	Moderate

Sylvaner

The major white variety of the Rheinpfalz and Rheinhessen regions of Germany, it is also grown elsewhere in central Europe and known sometimes as Österreicher or Franken. Both red and white grape clones are grown, particularly in Austria, though further plantings in that country are not encouraged. It ripens well before White Riesling, and its yield of wine is considerably higher. The Sylvaner wine has a pleasant ripe-grape aroma, low acidity (especially in older plantings) and a well-balanced, soft flavour; however it lacks the depth and complexity of Riesling. A slight harshness in the flavour of Sylvaner wine is typical of the variety, and often, in Germany, the wine is made sweeter for this reason.

Agronomy		
Soils	Likes deep, well-drained fertile soils
Vigour	Moderate
Pruning	Use medium to long canes
Disease susceptibility	Downy mildew—fair
	Powdery mildew—fair
	Bunch rot and *Botrytis*—fair to high
Wet weather	Susceptible to wet weather at harvest
Rootstocks	Use moderate or vigorous rootstocks such as Kober 5BB or 26G
Yields	Moderate to high

Lesser grape varieties grown in cool climates

There are many grapes grown in Europe which have generally lesser significance than the ones described above. Some are old varieties now declining in popularity. Some still have value in specific locations for specific markets, and some have importance as minor blending varieties. The following section contains a brief description of these varieties, most of which will have been mentioned previously in Chapters 1 and 2.

Aligoté Synonymous with Gigoudot blanc, Blanc de Troyés and Chaudenet gras. Aligoté is an ancient, white Burgundian grape, grown mostly in the Maconnais and Châlonnais districts alongside Pinot blanc and Chardonnay. In Burgundy the wine is sold as 'Bourgogne Aligoté. Reddish shoots, small berries in loose, cylindrical bunches plus low-moderate vigour are typical characteristics. The wine is of moderate quality and develops quickly.

Amigne and Arvine Two specialist grape varieties from the Valais region in Switzerland. Amigne, said to be the '*Vitis aminea*' of the Romans, produces a pleasant, perfumed wine. Arvine is considered to be of better quality, though neither has any significance outside its home district.

Auxerrois (Pinot Auxerrois) An ancient Burgundian white grape, thought sometimes to be a weaker-growing, earlier-maturing Chardonnay. It should not be confused with 'Gros Auxerrois' which is a synonym of Muscadet (Melon) and is a different grape variety. Auxerrois is a low yielding variety of good quality and many of its agronomic characteristics are identical to Chardonnay. It has been grown in Alsace and Southern German districts.

Blauer Portugieser A productive red grape of moderate quality from Germany, Austria, and Hungary. Its precise origin is not known, though it has been grown in Austria since the mid-nineteenth century. Vigorous in growth with a large leaf and cluster, it is very susceptible to powdery mildew and bunch rots. It is early ripening, and high yields plus a neutral wine character make it a popular parent for Austrian grape-breeding programmes and a number of useful crossings have resulted. Blauer Portugieser is best suited to the production of bulk, early-drinking red wine, or for blending with the more acid Pinot wines. Colour can be a problem in cold seasons.

Blaufrankischer A moderate to high quality red grape, although these factors depend on yield and site, soil type and season. The origin of this grape is not known and is presumed to be French, although it is grown in Austria and the Würtemberg region of Germany. It is a vigorous grower with upright-reddish shoots and a dark, tough leaf. The cluster is large and loose, thus more resistant to *Botrytis* and bunch rots. It is said to be very susceptible to poor weather at flowering and it sometimes fails to set even a moderate crop. Blaufran-kischer requires the best positions in cool climate districts as it is late ripening and retains high acid.

Bouvier An early ripening, moderately cropping, white grape from Austria. Muscat flavoured, the Bouvier wine has low acid and an aromatic bouquet. Yellow-green berries, with brown spots where exposed to sun, plus poor lignification of canes in marginal sites are typical.

Blauer Wildbacher An Austrian red grape of small significance which produces a soft wine for early drinking.

Bondola An Italian red grape, grown as a traditional grape in the Ticino district of Switzerland. Gives a rather harsh, tannic red wine, known there, when blended with other grapes, as 'Nostrano'. It is not grown in other cool climate regions; and even in Ticino it is being replaced by Merlot, thus it cannot be recommended.

Cabernet Franc An important, lesser grape of the Bordeaux region (especially St. Emilion) used for traditional clarets. Also known as Carmenet, Gros Bouchet, and Breton, Cabernet Franc is also grown in other French regions, notably in the Loire (for Rosé wine), and in the districts of the Midi where it produces a full-bodied, tannic wine of good quality. If compared to Cabernet Sauvignon, the wine is softer and more subtle, yet it retains the distinctive Cabernet aroma. Agronomic and ampelographic characteristics are similar to Cabernet Sauvignon.

Carmenére Listed sometimes as a minor red grape variety for the Bordeaux region, it closely resembles Cabernet Franc and is possibly the same grape. The name could be a corruption of Carmenet, one of Cabernet Franc's synonyms.

Elbling   An ancient, white grape cultivated in the German Mosel Valley and in Luxembourg since the Middle Ages. Possibly identical to the *Vitis alba* of the Romans, its name derives from 'albus', Latin for white. It is a high-yielding grape of low to moderate quality. Due to a low potential alcohol level and high acidity, it is most suited to the production of sparkling wines. In the Mosel district it is also known variously as Kleinberger, Alben, and Albig. A Red Elbling is also used in Mosel.

Gamay noir á jus blanc   The grape of the Beaujolais region, also known as Gamay noir Gamay de Beaujolais, Bourguignon noir and Petit Gamay. The Californian Gamay—the so-called Gamay de Beaujolais—turned out to be an upright clone of Pinot noir. The true Gamay of Beaujolais is grown in other regions of France, though in some, not under the most recommended conditions, for example in the Gironde and Alsace. It is also well known in Swiss plantings, notable in Valais. The characteristics include white, downy shoot tips, shiny leaves, and a compact, cylindrical bunch of round, coloured berries. This vine is sensitive to all moulds and a careful spray programme is required. Wine made in the Beaujolais style is consumed fresh, having great aromatic quality and a soft flavour (see the Maceration Carbonique method in Chapter 13).

Grüner Veltliner   A traditional Austrian grape, which is said to originate from Northern Italy. It is a vigorous grower, giving moderate to high yields of late-ripening grapes. Its demands for high soil fertility, lack of early spring frosts and a susceptibility to powdery and downy mildew, limit its usefulness for many districts. Under suitable conditions, this grape can prove a good variety for distinctive and fine quality white wines.

Humagne   An ancient grape type from the Swiss Valais region. It is not grown elsewhere and thus it cannot be recommended.

Heiden   A speciality grape in the Swiss Valais region, synonymous with Païen.

Limberger   A red grape variety from Austria, it is also grown in Germany, Hungary, and Yugoslavia. This grape appears to be identical to Blaufrankischer and could be just another synonym. It is also known as Lemberger and Blauer Limberger.

Muscat blanc á petits grains   A small-berried, ancient Greek variety of Muscat which spread to many European regions during the period of the Roman Empire. Known better as Muscat blanc, Muscat Canelli, Weiss Muscateller, and Muscuti, this is a vigorous grower, giving low to moderate yields of bunch-rot sensitive, tight bunches. Its susceptibility to both types of mildew, combined with a low yield and sensitivity to spring frosts, make this Muscat variety less attractive than the Muscat-Ottonel which replaced it in many cool climate regions.

Müllerrebe   Identical to Pinot meunier, see Major Varieties.

Muscadelle   Also known as Guillan musqué, Muscat fou, Guillan or Douzanelle, the Muscadelle grape derives it reputation from Bordeaux where it is used in the production of sweet or dry, blended whites. Muscadelle has characteristic yellow canes, darker at the nodes, large and loose bunches of muscat-flavoured berries and reddish shoots. The famous Constantia wine from the Cape Province in South Africa was said to have been made from Muscadelle. Despite its susceptibility to powdery mildew and *Botrytis*, it is recommended since its wine has a fine aromatic quality and the vine produces high yields.

Madelaine x Angevine 7672.   Not to be confused with Madelaine Angevine which is a table grape used in France. The early-ripening of this white grape has made it attractive to a number of grape growers in England.

Neuburger   Originally from the Niederösterreich region of Austria, Neuburger gives wine of good quality, but produces only a moderate yield. It is vigorous, having light-coloured canes with poor lignification. The leaves are three-lobed and small, and tight bunches of thick-skinned, red-green berries are typical. Fruit set problems, which result from inherent flower failure, provide a variability in yields; in the best sites these grapes produce fine white wine.

Petit Verdot   Also known as Verdot, this red grape is used to a small extent in the Bordeaux region for claret blends. Petit Verdot gives a dark, tannic and coarse wine of ordinary quality if tasted on its own: this grape is not grown in other cool-climate districts.

Pinot St Laurent   A vigorous low-yielding Austrian red grape with, at best, moderate quality. It has dark canes with short internodes and tough, dark and shiny leaves with large, well-filled bunches of acid-retaining berries. The wine is usually well coloured, but lacking depth in bouquet and has a coarse flavour. Poor fruit set and sensitivity to early spring frosts are also a characteristic of Pinot St Laurent. Further plantings of this grape cannot be recommended.

Röter Veltliner A traditional Austrian variety of uncertain origin. Also known as 'Malvoisie rose précoce', the Röter Veltliner is an early-ripening, vigorous variety which gives moderate yields of wine. The berry is light red-rose in colour and the variety has a clearly five-lobed leaf, the canes often show poor lignification. A large, shouldered and well-filled bunch is typical. The wine is low in acid, neutral in character, and matures fast.

Réze An ancient grape variety from the Valais district of Switzerland, associated with the 'Vin du Glacier', a traditional local wine style. A variety which is not known outside this district.

St Macaire Synonymous in the Bordeaux region with the Moustere and Bouton blanc; this is a lesser white grape used for the acid-fresh, 'fish wines' of the Entre-Deux-Mers district. Sensitive to both types of mildew, this variety is of low to moderate vigour. It has light-coloured canes, rose-coloured nodes, yellow-green leaves and small, loose clusters of acid-retaining berries.

Syrah Not a typical, cool-climate grape, Syrah is best known from the tannic, well-coloured and alcoholic red wine of the Rhône valley. These vines are also planted in the cool climate districts of Australia (as Shiraz) and, more surprisingly, in Switzerland as well; although, in the latter case, only to a limited extent and on the best sites. Syrah can be expected to ripen late, giving low yields of wine that retain high acidity; thus, in districts as cold as Switzerland it should be planted—if at all—in the most favourable positions. The vine has a yellow/green leaf, and is medium-sized and wavy-contorted: small berries in well-filled cylindrical bunches with oval berry shapes are typical. Syrah is sensitive to *Botrytis* and, if exposed to windy conditions, shoots can easily break, further reducing its low yields. Nevertheless, it is a high quality variety for the warmest sites.

Trollinger Synonymous with Blauer Trollinger, this is a vigorous red grape variety, grown mainly in the Würtemberg region of Germany. Large plantings can also be found in the South Tyrol district, from whence it is said to originate. In Austria it may be called Gross-Vernatsch, in England it is known as Black Hamburg, in France, Frankental noir, and in Italy it is named Schiavone. The large, attractive grapes of Trollinger ripen with high acidity and have, in cooler seasons, insufficient colour. Not a high-quality wine grape, but its attractive bunch and large berry make it useful as a table grape.

Welschriesling This is not a true Riesling, but another white grape grown in several important, cool-climate districts, notably in Austria. Known also as Riesling Italica and Wälschriesling, it produces high yields of late-ripening 'Riesling-like' wine with a floral, but earlier-developing style. The best quality wines come from late harvests, but even then they retain rather high acidity. Large, clearly five-lobed leaves and greenish tipped shoots are typical.

Veltliner-frührot An Austrian grape which in most aspects resembles, and probably is, Röter Veltliner or Malvoisie-rose-précoce.

Zierfandler A traditional Austrian white grape, known there also as Spätrot. Zierfandler is very much a speciality grape, producing a fresh wine for current consumption.

Rotgipfler An Austrian speciality grape, grown to produce full-bodied, flavoured red wine from late-harvested grapes near Gumpoldskirschen.

New varieties

The varieties which have so far been described are the established, classical wine grapes from the cooler districts of Europe. Not all the varieties are included, some like Grenache, Zinfandel, and Cinsaut have been omitted since they are not strictly cool-climate vines and, except in more favoured micro-climates, would not be grown.

Crosses between American species and *Vitis vinifera* — the 'hybrids', or 'Franco-American hybrids' — are not generally considered to be of sufficient merit to be included, although some will suit specific situations, and new breeding programmes are expected to produce better hybrids in the future which could have wider appeal.

There is, however, a further group that should be mentioned: these are the new vinifera crosses originating in Europe. These have been specifically bred to combine good agronomic characteristics with high yields, fine quality, and often, early-season cropping and cold tolerance. Some, like Kerner — described in the main section — are now rated as major varieties and many will be more significant than most of the lesser classical varieties. The following are a selection of the better-known new crosses.

Bacchus (White)
Nature of Cross:
(Sylvaner x Riesling) x Müller-Thurgau.
Origin:
Geilweilerhof Research Station, Rheinpfalz, Germany .
Crossing Number:
Gf 33-29-133.

A successful, patented new crossing which by 1981 was planted on 3174 hectares in Germany. It has many similarities to Müller-Thurgau; for example, in bud burst (early), ripeness date, and yield levels. It is suited to deep, fertile soils, well supplied with moisture. Recommended rootstocks are SO4 and 5C in good soils, 5BB and 125AA in poor soils. To produce the best quality, the harvest of Bacchus should be delayed some ten days after Müller-Thurgau and its fair resistance to bunch rots should allow this to be possible in most seasons. Bacchus wine has a distinctive, fruity bouquet and a grapey flavour together with a fair acidity at high levels of ripeness and a tendency to develop rapidly with bottle ageing.

Ehrenfelser (White)
Nature of Cross:
Riesling x Sylvaner.
Origin:
Geisenheim Institute, Rheingau, Germany.
Crossing Number:
Gm 9-93

One of the better crosses of Riesling x Sylvaner from the Geisenheim Institute. The area planted in Germany was 499 hectares by 1981. Ehrenfelser gives fair yields of mid to late ripening grapes, approximately two weeks before Riesling.

The wine is of Riesling quality and flavour and has good balance and a distinctive, flowery bouquet. Its improved sugar/acid ratio in lesser years makes it a more attractive grape than Riesling for less favourable sites. The variety is very resistant to bunch rot. High sugar levels of late harvest quality (Spätlese) are often recorded in German plantings.

Blauburger (Red)
Nature of Cross:
Blau Portugieser x Blaufränkischer.
Origin:
Klosterneuburg, Austria
Crossing Number:
ZNR 181/2.

A vigorous, heavy-cropping vine which produces bronze-tinged shoots, large leaves and well-filled large bunches — with tough-skinned, medium-sized berries — and heavy bloom. Blauburger produces well-coloured, tannic, but soft wine with a neutral bouquet which makes it a useful variety for blending with Pinot varieties, especially in lesser years. It ripens about the same time as Pinot noir or slightly earlier. Blauburger is highy susceptible to powdery mildew and, due to its well-filled bunches, it is moderately susceptible to *Botrytis* at harvest.

Breidecker (White)
Nature of Cross:
A complex crossing using Müller-Thurgau; one of its great grandparents was Siebel 7053.
Origin:
Geisenheim Institute, Rheingau, Germany.
Crossing Number:
Gm 4984.

Because of its small hybrid content, this grape cannot be planted in Germany. It has found a home in New Zealand where it was given the name Breidecker, after an early German grape grower. It has proved to be a very satisfactory alternative to Muller-Thurgau, as its early-maturing grapes show a fair degree of *Botrytis* resistance. Moderate in vigour and high yielding, it has soft, clearly three-lobed leaves and hint of pink colour in the ripe berries. It can be recommended for cool climate regions and for the production of light, fruity white wines for early consumption. The wine is clearly vinifera in character.

Faber - syn. Faberrebe (White)
Nature of Cross:
Pinot blanc x Müller-Thurgau.
Origin:
Alzey Research Station, Germany.
Crossing Number:
Az 10375.

Faber is a heavy-cropping grape which ripens with high sugar levels just after Müller-Thurgau. Bud burst is early, and it has a weak growth habit with the result that more vigorous rootstocks such as 5BB, 26G, or 125AA are usually recommended. Suited to most soils, its resistance to rots allows the grape to be used for late-harvest styles, giving an aromatic quality wine with full bouquet and flavour. In Germany there were 2115 hectares planted by 1981.

Morio Muscat (White)
Nature of Cross:
Sylvaner x Pinot blanc.
Origin:
Geilweilerhof Research Station, Rheinpfalz, Germany.
Crossing Number:
Gf 128-30.

A high-cropping, muscat-type grape which is noted for its rich bouquet. Morio Muscat ripens about ten days after Müller-Thurgau; bud burst is early, but fertile secondary shoots enable it to survive frost well. Grapes have lower sugar than Müller-Thurgau and the wine has a low alcohol content. However, if late picked, it is well balanced, full-flavoured, and aromatic. In Germany, there are 2975 hectares (1981), but most of the wine is blended and it is considered to be especially valuable in lesser vintages. The planting rate of this grape is now declining.

Pinotage (Red)
Nature of Cross:
Pinot noir x Cinsaut.
Origin: Stellenbosch University, Cape Province, South Africa.

This is not especially new since it was developed in the 1920s. The Cinsaut in South Africa is called Hermitage—hence the name Pinotage; the cross attempted to combine the yield of Cinsaut with the quality of Pinot noir.

Not strictly a cool-climate grape, Pinotage became for many years the most important, quality red wine produced in South Africa. There appears to be a wide clonal variation in Cape vineyards, the best wines giving a well-coloured, soft wine of distinctively raspberry-like aroma and tannic flavour. Significant plantings also exist in New Zealand, chiefly in the Henderson/Kumeu and Gisborne districts, although it is now declining in popularity. In cool climate areas, Pinotage ripens in mid to late season, before Cabernet, but after Pinot noir. The quality of the crop is largely dependent on the clone grown, on harvest condi-

tions, and on the level of virus infection. It has thin, dark and clearly five-lobed leaves with yellow-tipped shoots and an elongated, cylindrical bunch of oval berries. Pinotage fruit tends to be susceptible to wet weather at harvest when bunch rot can be a problem.

Perle (White)
Nature of Cross:
Gewürztraminer x Müller-Thurgau.
Origin:
Alzey Research Station.
Crossing Number:
Az 3961.

A variety which is particularly noted for its ability to withstand both winter chilling — due to good cane maturity—and spring frosts, due to late bud burst.

It is an early pinkish grape, ripening slightly after MüllerThurgau, giving wine, even in lesser vintages, with a flowery aroma, soft flavour, and low acidity.

This cross is susceptible to powdery mildew and *Botrytis,* the latter because of its tight bunch. Suitable rootstocks have been shown to include 125AA, 5BB, 5C, and SO4. Although approximately 350 hectares were planted in Germany (1981), it is no longer finding favour there.

Reichensteiner (White)
Nature of Cross:
Müller-Thurgau x Madelaine Angevine x Calabreser Fröhlich.
Origin:
Geisenheim Institute, Rheingau, Germany.
Crossing Number:
Gm 18-92.

Reichensteiner is one of the earliest-ripening grapes from the new crossing programmes. It matures before Müller-Thurgau and is particularly suited to cold areas where the season is short. It crops moderately well and good sugar/acid ratios are achieved, even in poor vintages. The wine has a pleasant, mild flavour without a pronounced bouquet. It is not widely planted in Germany, approximately 200 hectares by 1981. In England in 1988 Reichensteiner was the second most planted grape (after Müller-Thurgau)

Rieslaner (White)
Nature of Cross:
Riesling x Sylvaner.
Origin:
Alzey Research Station, Germany.
Crossing Number:
Az 11-17.

Due to its late ripening (slightly before Riesling) and low to moderate yield, the extent of German planting is low. Nevertheless, its exceptional quality and distinctive character should merit the attention of growers in other cool-climate districts where grapes of Riesling flourish well. The bouquet of Rieslaner wine is reminiscent of fresh peaches and apricots; its full flavour and fine balance — even in lesser years — develop a few days before, or at about the same time, as Riesling. It is suitable for blending or for late harvest styles and is recommended for better growing positions.

Scheurebe (White)
Nature of Cross: .
Sylvaner x Riesling.
Origin:
Alzey Research Station.
Crossing Number:
Az S 88.

A vigorous variety with late bud burst and yields comparable to Riesling. It ripens shortly before Riesling. Recommended rootstocks are SO4 and 5C; 5BB is unsuitable. Scheurebe has a powerful, peachy bouquet and pronounced flavour. In good years it produces outstanding wine and may be late harvested for Spätlesen or Auslesen styles; it is also useful for blending. 4083 hectares were planted in Germany by 1981 and the area has increased dramatically.

Schönburger (White)
Nature of Cross
Pinot Noir x IP1
Origin:
Geisenheim Institute, Rheingau, Germany.

Schönburger is popular in England due to its resistance to wet-weather diseases such as botrytis. This allows it to be picked later than other varieties, e.g. Müller Thurgau, which otherwise ripen at a similar time. It crops well and the wine has an attractive Muscat flavour.

Seyval Blanc (White)
Nature of Cross:
Seibel 4995 x Seibel 4986
Origin:
Bertille Seyve, France — called originally Seyve-Villard 5/276

Seyval Blanc is popular in cool climates such as England and Wales and has been used for several reasons; thus, although later than some other varieties, its disease resistance means it can hang on the vine late to get extra ripeness. Low vigour is useful in vigorous situations and

reliable cropping in cool climates is an added bonus.

Although a hybrid, some very good wines have been made; its neutral character make it a candidate for oak-ageing, methode champenoise or blending with more flavoured varieties.

Sieggerrebe (White)
Nature of Cross:
Gewürztraminer x Madelaine Angevine.
Origin:
Alzey Research Station, Germany.
Crossing Number:
Az 7977.

Although not widely planted in Germany (300 hectares by 1981), this is a useful early-maturing grape suited to cool climate areas such as Canada and Britain. It produces moderate yields of low acid wine with an intensive Muscat flavour. Vigorous in growth, this variety is suited to open-type training so as to avoid excessive shoot density and crop losses resulting from *Botrytis* and bunch rots. As with Gewürztraminer wet weather at flowering can reduce yields; reddish berries and short internodes are typical for both varieties.

Zweigeltblau (Red)
Nature of Cross:
Pinot St Laurent x Blaufränkischer.
Origin:
Klosterneuburg Institute, Austria.
Crossing Number:
ZNR 71.

A vigorous, upright-growing grape which produces high yields of medium-sized berries. Zweigeltblau ripens early with relatively low sugars and acids, producing a soft red wine suitable for early drinking or blending. The vine tends to produce a secondary set of fruit on sub-laterals which can lead to over-cropping and uneven ripeness at harvest. Usually spur pruned.

Zweigeltrebe (Red)
Nature of Cross:
Franken x Svatovavrinec.
Origin:
Klosterneuburg Institute, Austria.

Released for general cultivation in 1966, Zweigeltrebe has done well in a number of cool-climate districts in Europe. It is high yielding and early ripening, the wine being well coloured and having a fine aroma, a soft-tannic flavour and low acidity. Because of this character it is thought to be well suited for blending with the Pinot noir type wines when, in lesser years, they show greater acidity and poor colour. Its low resistance to mould disease may restrict plantings to low humidity, cooler areas of viticulture.

Some other new German varieties

Würzer (White) Ripening slightly after Müller Thurgau, it gives good yields of spicy aromatic wines with some potential.

Noblessa (White) Can produce fine wines. An early grape, but low yields and disease susceptibility give it limited potential.

Domina (Red) Reliable bearer of soft red wine with good market prospects. Yields are high, growth is vigorous. The grape ripens just before Pinot noir, it has good disease resistance and fair potential.

Nobling (White) While producing better wine than Sylvaner, it probably will not be able to compete with other new crosses. Susceptible to *Botrytis*.

Deckrot (Red) Mainly grown because of its high colour which makes it suitable for blending with Pinot noir in poor years. Not recommended as a varietal grape.

Kolor (Red) Similar comments to Deckrot. Both have very late bud burst.

Heroldrebe (Red) A reliable producer, but subject to *Botrytis*. Unlikely to be an important variety .

Helfensteiner (Red) Similar comments to Heroldrebe.

Osteiner (White) Some German literature suggests it could be better than Kerner or Ehrenfelser. The authors' experience in New Zealand does not confirm this and it appears to have little merit.

Rotberger (Red) of little merit except for Rose production.

A summary of the major characteristics of grape varieties

Further details of grape varieties are presented in Table 8.1 together with a summary of many of the factors which have already been mentioned. Vines are categorised according to vigour — low, medium and vigorous. A recommended pruning system is also presented.

If the plants are not grown on their own roots, then rootstocks are chosen which generally complement the vigour of the plant. Thus the vigorous Cabernet is complemented by a low-to-medium vigour rootstock. A vigorous rootstock has been recommended for the low-to-moderate vigour Pinot noir. This, however, is not always the case and cannot be regarded as a universal rule. For example, a weak rootstock has been suggested for low-vigour Gewürztraminer because it tends to set poorly if vigorous ones are used.

The date of bud burst is given because it can have significance in areas subject to late spring frosts. Naturally, those varieties with early bud burst are more at risk.

It is not possible to be too precise about the time of maturity; generally, varieties will ripen in the order indicated, but sometimes minor differences occur between districts due to soils, clones, and training methods.

Yields, likewise, will vary according to clones, districts, soils, training and pruning. As a general guide, a low yield might be considered to be in the range of 7-10 tonnes per hectare (2.8-4.0 tons per acre) and a high yield 17-20 tonnes per hectare (7-8 tons per acre).

Table 8.1 The major characteristics of cool-climate, grape varieties

Variety	Preferred position & soil types	Vigour of vine	Rootstock vigour preferred	Recommended pruning system	Time of bud burst	Time of maturity	Expected yields
Bacchus	Not demanding, but not too dry	Moderate	Moderate to vigorous	Head cane	Early	Early	High
Cabernet Sauvignon	Not demanding, but needs well-drained soils	Vigorous	Low to medium	Cordon cane	Late	Late	Low to moderate
Chardonnay	Well-drained, fertile and cal careous soils	Moderate	Vigorous	Head cane	Early	Early to mid season	Low
Chasselas	Not demanding	Low to Moderate	Moderate	Head cane	Late	Early	Moderate to high
Chenin blanc	Not demanding	Moderate to vigorous	Moderate to vigorous	Cordon cane	Mid season to late	Late	Moderate to high
Ehrenfelser	Not demanding	Moderate	Moderate	Head cane	Early	Mid season	Moderate
Faber	Not demanding	Moderate	Vigorous	Head cane	Early to mid season	Early	Moderate to high
Gewürztraminer	Deep soils with low nitrogen	Low to moderate	Low	Cordon cane	Early	Early	Low to moderate
Kerner	Well drained- but not excessively dry soils	Moderate	Vigorous	Head cane	Mid season	Mid season.	High
Malbec	Well-drained, fertile soils	Moderate	Low to moderate	Cordon cane	Mid season	Mid to late season	Moderate
Merlot	Deep and fertile soils	Moderate	Moderate	Cordon cane	Late	Late	Moderate
Meunier	Well-drained and calcareous soils	Moderate	Vigorous	Head cane	Mid season to late	Early	Moderate
Morio — Muscat	Deep fertile	Low to moderate	Vigorous	Head cane	Early	Early to mid season	High
Müller Thurgau	Deep, fertile and not excessively dry soils	Low to moderate	Vigorous	Head cane	Early to mid season	Early	High
Muscat Ottonel	Deep, fertile and not excessively dry soils	Low	Moderate	Head cane	Mid season	Early	Moderate
Perle	Well-drained soils	Low to moderate	Moderate to vigorous	Head cane	Late	Early	Moderate
Pinot blanc	Well-drained, stony or calcareous, not excessively fertile soils	Moderate	Vigorous	Head cane	Early	Early	Moderate

Continued overleaf

Table 8.1 Continued

Variety	Preferred position & soil types	Vigour of vine	Rootstock vigour preferred	Recommended pruning system	Time of bud burst	Time of maturity	Expected yields
Pinot gris	Not demanding	Moderate	Vigorous	Head cane	Mid season	Mid season	Moderate to high
Pinot noir	Well-drained, deep and calcareous soils	Low to moderate	Vigorous	Head cane	Mid season	Early to mid season	Low to moderate
Reichensteiner	Not demanding	Moderate	Moderate to vigorous	Head cane	Early	Early	Moderate
Rieslaner	Well-drained, stony, not fertile soils	Moderate	Low to moderate	Head cane	Mid season	Late	Low to moderate
Riesling	Well-drained, stony, not excessively fertile soils	Moderate to vigorous	Low	Head cane	Late	Late	Low to moderate
Sauvignon	Well-drained, less fertile soils	Vigorous	Low	Cordon cane	Mid season to late	Late	Low to moderate
Scheurebe	Fertile, not too dry soils	Moderate to vigorous	Moderate	Head cane	Late	Mid season	Moderate
Sémillon	Well-drained, not excessively fertile soils	Moderate	Low to moderate	Cordon cane	Late	Late	Moderate
Siegerebe	Not demanding	Moderate to vigorous	Low to moderate	Head cane	Mid season	Late	Moderate to high
Sylvaner	Not demanding	Moderate	Moderate to vigorous	Head cane	Mid season	Mid season	Moderate to high
Zweigeltrebe	Deep fertile soils, not too dry	Moderate to vigorous	Vigorous	Cordon cane	Mid season	Early	High

Rootstocks

The importance of rootstocks was described in Chapter 6. The following list is a guide to the major rootstocks for *vinifera* grapes and may be used in conjunction with the summary shown in Table 8.2.

A.R.G.l, AxR I or Ganzin 1 *(Aramon x Rupestris Ganzin 1)* Vigorous growth but less resistant to *Phylloxera* and nematodes. A rootstock which has a favourable effect on set and yields and, for most varieties, is suitable for medium to heavy soils. Unsuitable for calcareous soils, or *Phylloxera*- or nematode-infested areas.

Couderc 3309 *(Vitis riparia x Vitis rupestris)* Produces vines of low to moderate vigour in deep fertile loams well supplied with moisture; unsuited to dry and shallow soil conditions. Good resistance to *Phylloxera* and moderate to nematodes and lime-induced chlorosis. Has tendency to induce potassium deficiency in overcropped young vines on clay soils. Recent Californian experience indicates the need for good nutrition for young vines grafted on 3309C. Good for varieties with poor set but due to high fruitfulness crop removal is important in young vineyards. Grafts easily.

Fercal *(Vitis berlandieri x Colombard)* **x 33EM** A new French rootstock bred for tolerance to calcareous soils, suitable for well-drained but not shallow soils. Prone to magnesium deficiency. Not a very productive rootstock, but one which may advance maturity compared with more vigorous stocks.

Table 8.2 Major rootstock characteristics

Rootstock	Preferred position and soil	Vigour of growth	Phylloxera resistance	Other features
A.R.G.-1	Suited to heavier soil types. Low tolerance to calcareous soils and very dry positions	Vigorous	Low	Vigorous rootstock with long vegetative cycle
Couderc 3309	Deep fertile soils, not dry or shallow conditions	Low-moderate	Good	Will improve fruit set , do not overcrop in early years
Fercal	Suits well drained but not shallow soils; tolerant to calcareous soils	Moderate	Good	Can advance maturity, light cropping, prone to magnesium deficiency
Gravesac	Well drained soils of low fertility, tolerant to acid soils	Moderate-vigorous	Good	Behaves rather like SO-4
Kober 5BB	Suitable for calcareous or clay soils, but not excessively dry positions	Vigorous	Very good	Relatively short vegetative season; vigorous, early-season growth typical
Kober 125AA	Greater drought resistance than Kober 5BB, suited to poor, stony soils	Very vigorous	Good	Used for special conditions only
Portalis	Deep, fertile soils, not suitable for dry positions or calcareous soils	Low-moderate	Very good	Thickening below graft-union typical
Richter 110	Not demanding, has good resistance to drought, good for slopes	Vigorous	Very good	Recommended ahead of other Richter stocks e.g. R-37, R-44, or R-99
Rupestris St George	Lighter soil types, well-drained positions; good resistance to drought	Vigorous	Good	Tendency to produce shoots below graft-union
Schwarzmann	Not demanding, but not suitable for excessively heavy soils	Moderate	Good	—
SO-4	Not demanding, but likes light, well-drained soils of low fertility	Moderate-Vigorous	Very good	Resistant to nematodes and drought. Especially suited for grape varieties with poor set
143A	Deep and fertile soils, not too dry. Low tolerance to calcareous soils	Low-moderate	Low	Used for vigorous varieties with irregular set
5C Teleki	Suitable to well-drained, fertile soils, moderate drought resistance High tolerance to calcareous soils	Moderate	Good	Original material appears variable in some characteristics and improved clones from Geisenheim are recommended (e g Geisenheim No 6 and No. 10)
26-G	Best for well-drained soils, but not too demanding	Low-moderate	Low	Suited for varieties with poor set
101-14 Millardet	Suited to deep soils of low lime content, not drought tolerant	Low-moderate	Good	Useful for advancing maturity of late varieties

Gravesac (16149 Couderc x 3309) Another new rootstock bred in Bordeaux specifically for tolerance of acid soils. It is moderately vigorous and gives good yields. In Bordeaux it performs like SO-4.

Kober 5BB *(Vitis berlandieri x Vitis riparia)* Selected as one of the best rootstocks for calcareous soils. Vigorous in growth with a relatively short, vegetative season and a good affinity with most varieties: it appears to be superior to similar hybrids from France. Kober 5BB is popular in all areas of northern Europe, however, its vigorous growth in the early season can cause problems. higher-yielding grape varieties show symptoms of potassium deficiency on this stock and, with more vigorous varieties, it can cause imperfect set. It is less suitable for positions of prolonged drought and those affected by severe winter frosts—the root will tolerate -8°C (18 °F) at 30 cm (12 in). Recommended for varieties with moderate vigour and yield. Good resistance to *Pyhlloxera* and nematodes.

Kober 125AA (*Vitis berlandieri x Vitis riparia*) In most aspects similar to Kober 5BB, though more vigorous in growth and is recommended for special conditions only. Best suited to poor soils with high-yielding varieties, such as Müller-Thurgau.

Portalis or Gloire de Montpellier (*Vitis riparia*) Portalis is a clonal selection of the American species *Vitis riparia*. It has a good resistance to *Phylloxera*, but is not very suitable in drought conditions or for calcareous soils. Good for deep fertile soils, it is of low vigour and is recommended for more vigorous varieties to make growth more manageable. Excess thickening below the graft-union is characteristic. Portalis can improve flowering and set and, in some cases, increase the berry size and bring forward harvest maturity.

Richter 110 (*Vitis berlandieri x Vitis rupestris*) Vigorous, has a good affinity with most varieties; resistant to drought conditions. Richter 99 has good resistance to *Phylloxera* but only moderate resistance to nematodes— it has a favourable effect on set and yields. Generally, recommended ahead of other Richter stocks.

Rupestris St. George Also known as Rupestris du Lot, this rootstock, for most varieties, is well suited to medium and lighter soils. It is resistant to *Phylloxera* and drought conditions and has a tendency to produce shoots from below the graft. Growth is vigorous and this rootstock is less suitable for varieties with irregular set.

Schwarzmann (*Vitis riparia x Vitis rupestris*) A selection which has a better resistance to drought, sandy soils and calcareous soils than Portalis. Moderate in vigour and good for more vigorous varieties and lighter, warmer soils.

SO-4 (*Vitis berlandieri x Vitis riparia*) A selection originating in Oppenheim. Its recent increase in popularity, in both France and Germany, is due to the beneficial effect it has on the maturation of canes, set and yield, and its suitability to lighter, poor soils with high stone content. Vigorous and not recommended for very dry conditions. Recommended for varieties with irregular set. Resistant to *Phylloxera* and nematodes.

143A (*Aramon x Vitis riparia*) A less-vigorous rootstock, also known as Aripa, which possesses a good affinity with most varieties and has a favourable effect on yields. It is suitable for fertile soils and more vigorous varieties— especially those of irregular set. Lower tolerance of *Phylloxera* and calcareous soils.

5C Teleki (*Vitis berlandieri x Vitis riparia*) Like Kober 5BB this rootstock was selected for its tolerance to calcareous soils. It has many of the attributes of 5BB except that it tends to bring ripeness forward. It therefore has special value for very cool climates.

26G (*Trollinger x Vitis riparia*) Originally from Geisenheim, it is regarded as possibly the best rootstock for Rhine Riesling in Germany, especially in the Moselle. Recently it has been extensively used in Austria for Grüner Veltliner. The slow growth during flowering and set is said to have a favourable effect on yield and bunch development. 26G is tolerant of calcareous soils and is best suited to medium-light, well-drained, lighter soils. Drought resistance moderate, *Phylloxera* resistance low, use with caution in *Phylloxera*-infested areas.

101-14 Millardet (*Vitis riparia x Vitis rupestris*) Gives vines of low to moderate vigour and is suitable for deep soils of moderate to low lime content. Not appropriate for dry and well-drained positions on slopes, The rootstock has good resistance to *Phylloxera* and is moderately so to nematodes. It grafts easily and the characteristically short vegetative life cycle is valuable for advancing maturity of late varieties (e.g. Cabernet Sauvignon) in cool climates.

CHAPTER 9
PESTS AND DISEASES OF GRAPE VINES

A large number of fungi, bacteria, viruses, insects and nematodes attack grapevines. Some cause serious problems, some only minor ones, some are prevalent in one area and not in another, some are effectively universal.

The descriptions which follow are not intended as a detailed account of all the pests and diseases which can affect vines, but will cover those disorders which are found in most districts. A few of the more common and widely-used sprays will be mentioned.

Growers wishing to have pests and diseases identified, or be given spray programmes to control them, should consult local advisory officers or chemical spray representatives. This sort of assistance is normally not difficult to obtain.

Pests

Phylloxera

This is a pear or oval-shaped root-aphid which spends its life living on the roots of grape vines. The microscopic insects are yellowish-green, or yellowish-brown, and produce bulbous yellow swellings on older roots. Above the ground infected vines can often be recognised by poor growth and yellowish leaves. Adjacent vines can become infected via roots, or by the insect crawling short distances over the ground. Infestation over greater distances can occur in the transfer of infected cuttings or grafts, or on soil carried by tools and implements. While the insect spends most of its life history below the ground, a 'crawler' form will sometimes venture onto the leaves. This can be transferred to neighbouring vineyards by strong winds, or on mechanical harvesters. A winged form occurs on American *Vitis* species.

Where *Phylloxera* is not present in a district, vines can be grown on their own roots. This has clear advantages, since cuttings are easy to propagate and growth of the vine is normally very satisfactory. However, under these conditions it is very important that no infected material is introduced; cuttings or other material from infected areas should be washed and fumigated, or dipped in emulsions of insecticide before leaving the infected district. In most countries where *Phylloxera*-free areas exist there are specified treatments which must be applied before grape material can be introduced. If vines are suspected of having *Phylloxera* in a vineyard where infection has not previously been recorded, the authorities should be notified and, if infection is confirmed, the vines should be removed and the soil sterilised, hopefully to prevent a further spread in the vineyard, before replanting with grafted vines. No chemical treatments can satisfactorily control this pest in the vineyard.

Nematodes

Nematodes are microscopic worm-shaped organisms which, like *Phylloxera*, feed on the roots of vines. Most soils contain some nematodes, but not all districts have problems; either because the species are not damaging to grapes or because the conditions are not suitable for their growth. Expert advice is usually needed to identify the damage caused by nematodes which, generally, results in weakened growth and lowered production. Gall-like swellings are sometimes found on the roots of infected vines. In areas where nematodes are known to be a problem, fumigants are sometimes applied to the soil before planting. Again, expert advice is needed before embarking on such a programme. Rootstocks vary in their resistance to nematodes and careful selection is needed in areas where the problem is encountered.

Rootstock resistance to nematodes and *Phylloxera* is described in Chapter 8.

Other pests

Many other pests attack grapes. They include chewing and sucking insects and the damage caused can vary from slight to severe. Because they vary considerably from place to place they will not be discussed here and local knowledge should be sought for identification and control.

Diseases

Downy mildew (*Plasmopara viticola*)

This is a disease which is common in moist weather and less common in dry conditions. Discoloured patches appear on the leaves and, in severe cases, the underside of the leaf becomes covered with dense white down. Berries attacked post-bloom may wither and fall. Spray when shoots are 2-3cm long with copper compounds (Bordeaux mixture, copper oxychloride, 'Kocide' or 'Champion') then, if required, spray at 10-14 day intervals until blossom with copper or other materials such as metiram or mancozeb. Sprays used for *Botrytis* at blossom will also control downy mildew. Renew spraying with previous materials till beginning of ripening.

Blackspot (*Anthracnose*)

Like downy mildew this is a wet-weather disease which shows as black spots developing on the growing leaves. Sprays used for downy mildew will control black spot.

Dead Arm (*Phomopsis*)

Another wet-weather disease appearing as dead areas on cordons. Use same sprays as for blackspot and downy mildew.

Botrytis or Grey Mould (*Botrytis cinerea*)

This is probably the most serious rot in wet weather. In moist climates a regular spray programme must be adopted since first infection may occur at blossom-time, even if it then lies dormant until maturity. In dryer climates it is less of a problem and one or two sprays may be

Figure 9.1 *Botrytis* . The tight bunches of Riesling are particularly susceptible to *Botrytis* and it is not a suitable variety for areas with warm wet weather

sufficient. Use 'Euparen' ('Elvaron') or 'Shirlan' at capfall to fruit set (1-2 sprays), apply same material just before bunch closure. Post veraison, use up to three sprays of 'Rovral', 'Ronilan' or 'Sumisclex'.* Training methods which expose the grapes to sun and rain tend to increase sunburn, bird damage and splitting, and this predisposes the grape to *Botrytis* at harvest. On other systems, where a mass of foliage surrounds the berries and leads to a build up of humidity, *Botrytis* can also become serious. The upright systems described in this book give a reasonable protection from rain and birds. As long as the density of foliage is controlled — if necessary by trimming or even limited leaf-removal near maturity—air flow is encouraged, humidity is discouraged and disease reduced.

Bunch rots

These are caused by fungi such as *Diplodia*, *Aspergillus*, and *Cladosporium* and, in contrast to *Botrytis*, they cause the berries to have a pungent, vinegar odour — thus their other common name 'Sour rots'. They tend to attack berries which have already been physically dam-

---

* Rotational use of different sprays is known to slow down the build up of resistance, and this procedure is recommended .

aged or infection with *Botrytis* has already occured. Similar control measures to *Botrytis* are needed.

Powdery mildew or *Öidium (Uncinula necator)*

American grapes have a good resistance to this disease but European grapes are much more susceptible. In dry climates, where downy mildew, *Botrytis* and other rots cause few problems, *öidium* is the major disease. It causes a dust-like, grey to white coating on the leaves and fruits which are prevented from ripening and may split. Berries that have begun to ripen, or are already ripe, are not affected. Sulphur sprays or dusts are standard protectant sprays and may start when shoots are 20cm long and continue until veraison at 2-week intervals. If infection has started, use eradicants such as 'Rubigan', 'Alto', 'Bayleton', 'Systhane', then return to sulphur.

Collar rots and Crown gall

In heavy soils, especially in wet seasons, rots can occur around the base of the trunk which will quickly kill the plant. These can be due to a range of fungi of which, perhaps, the most common are species of *Phytophthora*. This is mainly a problem with young vines. Physical damage to the trunk should be avoided and soil and weeds should not be allowed to accumulate round the base of the trunk. If collar rots occur, clear around the base of the trunk so that it is exposed to both sun and air. Some growers apply lime sulphur to the infection with fair success.

Crown gall is also an infection around the base of the truck of vines caused initially by physical damage to the trunk. In cold climates this may be due to freeze damage at −15 to −20°C.

Collar rots and crown gall cannot easily be controlled once infection has begun, but cleaning of the wound and exposure to air may assist recovery.

Viruses

There are a number of virus diseases; one such is 'Leaf Roll', which cause premature autumn colouring in red grapes and the edges of the leaves to roll under; another is 'Fan Leaf', which distorts the leaf so it has the shape of a fan. Other viruses are less conspicuous and some may be difficult to detect, yet may still have undesirable effects.

A virus seldom kills the plant and the effects of one or more are usually to reduce vigour, reduce the crop or delay ripening due to an overall reduction in the efficiency of the vine. No sprays or vineyard treatments will control a virus infection, but there are ways of reducing or eliminating viruses. There are two methods; the first is the process of clonal selection, whereby growers and nurserymen select clean, healthy, heavy-cropping scion and rootstock material for propagation. This will tend to eliminate those plants severely infected with virus. The second method is that of heat treatment and virus indexing. Young vines are grown under artificially-high temperatures and the vigorously-growing tips are taken off for further propagation. These are then indexed. This means that they are grafted to other non-infected vines which are particularly sensitive to certain viruses and quickly show symptoms of infection. If the heat treatments have been successful, the virus will have been eliminated and the progeny from that material is labelled 'free from known virus' (FKV) or, incorrectly, 'virus-free'. The resulting vine is usually more vigorous, gives better yields, ripens better and red grapes have a deeper colour. Heat treatment and clonal selection together give the best results.

The spread of a virus is usually very slow, although in some districts certain species of nematodes will carry viruses quite rapidly from one plant to another. An FKV variety should never be grafted onto a rootstock which has a virus infection since it will quickly travel into the scion. Likewise, an infected scion should not be grafted or budded onto FKV rootstock.

Birds

Birds can be a considerable problem, especially in small vineyards. They peck at a large number of berries and eat a few from veraison onwards. Those berries that are pecked succumb to *Botrytis* and other rots, and the juice may be sucked out by wasps and flies. Larger vineyards, where the total area is over 50 hectares (120 acres), suffer much less from bird damage. Birds are territorial by nature and the elimination of nesting sites can go a long way to reducing their population density.

Some trees carry more nests than others and it is advisable to restrict planting to those species which attract fewer nests. Some growers try to eliminate bird infestations by the use of poison but it is not always acceptable and can be dangerous. Shooting to kill or scare away birds is also used near harvest and automatic bangers and other devices can be effective for limited periods. The best control of birds is by a range of scaring devices which are varied over time. Once birds become habituated to such a device it should be removed and replaced by another. Netting is sometimes used on small plots and kites and other ingenious devices are used.

Physiological

Fruit set refers to the successful transfer of pollen from the male parts of the flower (anthers) to the female (style and stigma)—called *pollination*; the successful fusion of male and female is *fertilization* and both are prerequisites to a good set and satisfactory yields. Cold wet weather over the flowering period may reduce set and considerable variation between varieties and clones occurs. Many which set well in warm dry climates are notoriously poor in cooler moist areas. Gewürztraminer, Merlot, Malbec are good examples. Methods to improve set have been discussed in Chapter 3.

Early bunch-stem necrosis (EBSN) attacks the bunch from the time it is 2cm long until capfall. While the end result is the same as poor set (i.e. low crops) its manifestation is different. EBSN causes sections of the bunch to shrivel and dry so the bunch ends up with fewer branches. Characteristically the dead portions often remain attached. EBSN is exacerbated by stress factors such as, poor nutrition, drought prior to capfall and severe shade around the bunches; leaf loss due to say, severe hail or disease, will increase the disorder and ethephon sprays prior to capfall induce EBSN. Cool overcast and wet weather will also enhance the disorder. The grower can clearly modify conditions to reduce stress, but the most important measure is to reduce shade by good foliage management — for discussion of foliage management see Chapter 7.

Bunch-stem necrosis (BSN) — also called 'waterberry' and 'shanking' — is a similar but later manifestation. The stems of the bunch become necrotic and berries cease growing and shrivel. They may remain attached or fall off. Some evidence suggests that poor weather near flowering may also predispose the vines to this later disorder. Magnesium sulphate at 2.5 kg/100 litres sprayed at veraison and 7-10 days later will sometimes reduce it.

Problems with **splitting** have already been mentioned. This is a physiological response to excessive moisture at ripening. The grape absorbs large amounts of water and, as the skin will not expand sufficiently to take the extra volume, it splits. Control is impossible, but training systems which keep excessive water off the berries or allow quick drying can help.

Top Photo: Poor set in Cabernet Sauvignon. The large berries will have a full complement of seeds, the smaller ones will have a reduced number or none at all. The very small ones (coloured green) are not fertilized and normally fall off; sometimes, as here, they remain attached but they do not ripen.

Lower Photo: Early bunch-stem necrosis (EBSN). The withered sections of the bunch were infected prior to capfall.

CHAPTER 10
VINE GROWTH—THE ANNUAL WORK PROGRAMME

Advanced bud swell

As buds swell they become more sensitive to frost. The soil by now should be less moist and cultivation, if it is being used, should have begun about ten days before this stage to allow compaction which will help to reduce danger from Frost. Weeds are rotary-hoed, or disced between rows and hand-hoed, or sprayed within the row with a knock-down spray and pre-emergence herbicide. Lime sulphur should be applied at this stage as a clean-up spray for scale insects, some mites, mildews and anthracose (black spot).

Bud burst

Buds burst later than on most stone fruit such as peaches, apricots and plums, and at about the same time as apples and pears. From this stage onwards the shoots are susceptible to frost. In areas susceptible to wet weather diseases (*Botrytis*, downy mildew, anthracnose) copper sprays are applied, these include Bordeaux mixture, copper oxychloride, 'Kocide', 'Champion'. The latter two may reduce frost damage when sprayed over the bud-burst period.

Early shoot growth

For eight to ten weeks after bud burst the shoots will rapidly elongate. During this time the grower should begin 'tucking-in' the shoots between the wires of the trellis. Strict attention to this matter is most important, since grapes on vines tucked in late become heavily shaded and ripen later than normal, also spray penetration is poor and humidity is high. By flowering time two sulphur sprays or dusts should have been applied. Use eradicant fungicides (e.g. 'Rubigan', 'Systhane') if powdery mildew has become established.

Flowering (capfall)

Flowering will occur about eight weeks after bud burst. Weed control should be continuing. Do not hoe where pre-emergence sprays have been used until as late as possible; breaking the soil surface will destroy their effectiveness. The vines may have reached the top of the trellis by this stage. If topping is needed, now is a good time to do it since it might assist fruit-set. Pray for fine weather! Vines do not depend on insects for pollination, nevertheless, do not spray flowers with insecticide or neighbouring beekeepers will be among your enemies. Immediately after flowering begin spraying, if required, for insects such as leaf roller (*Tortrix*). A *Botrytis* spray is essential at flowering and again before bunch closure. Use 'Euparen' ('Elvaron') or 'Shirlan'.

Fruit-set to véraison

Growth of berries will be rapid for four to six weeks and then slow down. Side shoots will be conspicuous now and some lateral trimming to stop the vine becoming too dense may be required. Continue weed control activities; spray for mildews etc if needed.

Véraison

Véraison will occur eight to ten weeks after blossoming. Except in moist wet weather, shoot growth will slow down now and further topping and trimming are not usually required. The limited removal of leaves around bunches will help air circulation, assist drying and make maximum use of the heat from the sun which will begin to get weaker as the days go by. Further infections of powdery mildew will not occur but, as the sugar rises, *Botrytis* will become a problem in moist humid conditions. Use compounds such as 'Sumiclex', 'Rovral', 'Ronilan' according to the manufacturers instructions. Keep an eye on birds and use protective measures shortly after véraison, preferably before the first signs of damage are noticed. Begin to take weekly acid and sugar measurements: taste the berries to monitor flavour development.

Harvest

The busiest time of the year and the most worrying. Where the weather is deteriorating and the grapes are still not fully ripe, the grower and the winemaker must decide whether to risk considerable crop loss if picking is delayed, or lower quality if picking occurs early.

Post-harvest

Relax. Let weeds grow if green cover is needed over winter. Use a clean-up spray of lime sulphur or Bordeaux mixture, winter oil and insecticides if required for downy mildew, black spot and mealy bug.

Leaf fall

Pruning can begin and continue during the winter months, unless late pruning in frost-prone districts is required. Remove prunings and burn, or chop up and incorporate in the soil. Check trellis, machinery, etc and prepare a work programme for the coming season. Enjoy mulled wine by a log fire.

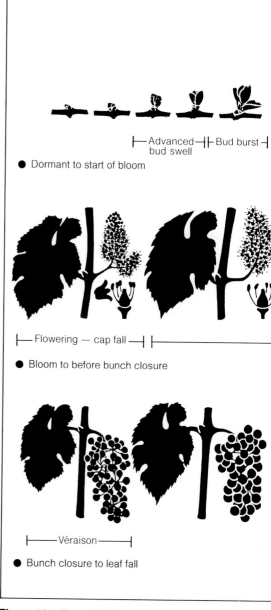

Figure 10.1 Stages in vine growth (Diagrams courtesy of Fruitfed Ltd, NZ)

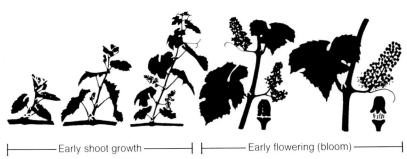

├────── Early shoot growth ──────┤ ├────── Early flowering (bloom) ──────┤

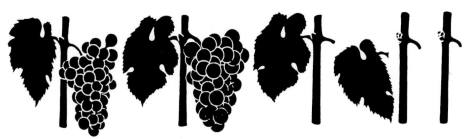

├────── Late flowering ──────┤

├──── Harvest ────┤ ├──── Leaf fall ────┤ ├── Dormant ──┤

REFERENCES AND FURTHER READING

Anon. 1982. Grape Pest Management, *Univ. California Publ. No. 4105*, Berkeley.

Champagnol, F. 1984. *Eléménts de Physiologie de la Vigne et de Viticulture Generale.* F. Champagnol, Saint-Gely-du-Fesc.

Coombe, B.G., Dry, P.R. *Viticulture. Volume 2, Practices*, Winetitles, Adelaide.

Galet, P. 1983. *Précis de Viticulture*, Déhan, Montpellier.

Galet, P. (translated Morton, L.T.) 1979. *A Practical Ampelography*, Cornell University Press, Ithaca & London.

Pongracz, D.P. 1978. *Practical Viticulture*, David Philip, Cape Town.

Pongracz, D.P. 1983. *Rootstocks for Grape Vines*, David Philip, Cape Town.

Smart, R.E., Robinson, M.D. 1991. *Sunlight into Wine : A Handbook for Winegrape Canopy Management*, Winetitles, Adelaide.

Vogt, E., Götz, B. 1977. *Weinbau*, Ulmer Stuttgart.

Weaver, R.J. 1976. *Grape Growing*, John Wiley & Sons, New York.

Winkler, A.J., Cook, J.A., Kliewer, W.M., Lider, L.A. 1974. *General Viticulture*, Univ. of Calif. Press, Berkeley.

PART III

WINEMAKING

INTRODUCTION

There is one biochemical reaction which is the basis of all winemaking operations. This is the conversion of sugar to alcohol and can be represented as follows:

$$C_6H_{12}O_6 \xrightarrow{\text{yeast}} 2C_2H_5OH + 2CO_2 + \text{energy (56 Kcal/mole)}$$

i e sugar $\longrightarrow$ ethyl alcohol + carbon dioxide + energy (heat)

In other words, if a solution which contains a sugar such as fructose or glucose is allowed to ferment, it will produce alcohol and carbon dioxide gas and, at the same time, it will release a certain amount of energy—in this case heat energy.

In Part III the materials involved in this reaction will be reviewed and the basic principles behind the process identified. Chapters on white, sparkling and red wine production are followed by a description of small-scale winemaking, a process valuable for the research and evaluation of new varieties.

The final chapters consider the wine itself, how it can be matured, stored and evaluated, and take a look at the future of winemaking in cool climates.

Analysing wine in a small winemaking laboratory

Note: Some readers may find it easier to read Chapters 12 and 13 — the production of white and red wines — before studying Chapter 11, the Principles of Winemaking. This will be particularly valuable if the reader is not generally familiar with the stages of winemaking: Figure 11.5 will also be helpful in this respect.

CHAPTER 11

THE PRINCIPLES OF WINEMAKING

The materials involved in fermentation

Carbohydrates

The sugar shown on the left-hand side of the biochemical reaction which introduces Part III is a member of the class of compounds known as carbohydrates. They are formed by the action of sunlight on the chlorophyll contained in the green leaves of plants, using water in the plant and carbon dioxide from the air as raw materials. Effectively, carbohydrates store the energy from the sun in a form which can be used by the plant, or by an animal which eats the plant. These stores, usually in the form of starch in the plant, or fat in the animal, can in turn be broken down at the appropriate time to provide energy and release carbon dioxide. Carbon dioxide is a waste product but the energy is necessary for growth and movement.

Unlike a number of other fruits, grape berries contain only sugars and no starch. The simplest sugars are glucose and fructose which are converted by yeasts to alcohol. More complex sugars like sucrose—normal table sugar—can be fermented by yeasts, although they need first to be converted by the yeast enzymes to glucose and fructose. Some other types of sugars cannot be fermented and may still be present in very small quantities in the wine after fermentation. The grape is an ideal base for winemaking because it consists entirely of glucose and fructose, and because the levels are so high it is usually not necessary to add further sugar to produce satisfactory alcohol levels. The other advantages of the grape will be discussed later.

Alcohol

There are many types of alcohol some of which, like methyl alcohol which is present in methylated spirits, are poisonous. Ethyl alcohol (wine alcohol) itself is poisonous if taken at too high a level but, in moderation, it has effects on the body which many regard as therapeutic. Alcohol, of course, is essential for any wine and a concentration which is too low makes the wine appear thin and lacking in body. Table wine contains between 8 and 14 per cent ethyl alcohol which is the result of fermenting a solution containing between 16 and 25 per cent sugar. Also produced are very small quantities of other more complex alcohols, and these add to the distinctive quality and flavour of the finished wine.

Carbon dioxide

The release of carbon dioxide gas from a liquid indicates that the yeasts are still active and that fermentation has not yet finished. The bubbles produced on opening a bottle of Champagne, however, occur because, when the wine was bottled, additional fermentation was induced in the bottle by adding a small quantity of sugar and yeast. Since gas was not able to escape, it remained in the liquid till such time as the cork was removed and the release of pressure allowed the gas to come out of the solution (see Chapter 12). Carbon dioxide is a waste product of fermentation, but its presence has some very desirable effects as will be seen later.

Energy

The energy released during the process of fermentation is available for the yeasts and assists their growth; however only about 40 per cent is so used and the remainder is given off as heat—this is why the temperature of the liquid rises during fermentation. In small containers this may not be noticeable since heat is quickly lost but, in larger ones, temperature rise might be considerable and cooling may be necessary.

Yeasts

The organisms which are responsible for fermentation are microscopic, single-celled yeasts. They are plentiful in nature and can be carried by air currents to almost any location where fermentable materials are likely to occur. There are numerous types of yeast, some of which are wine yeasts, and others baker's or brewer's yeasts; all have a similar capacity for using sugar as a source of energy.

One of the significant things about yeasts is their ability to use sugar for energy without requiring oxygen. This is not common; most organisms need oxygen before they can use a carbohydrate as an energy source. On looking at the following formula, and comparing it to that of fermentation given in the Introduction, it will be seen that no alcohol is formed and that the amount of energy released is much higher.

$$C_6H_{12}O_6 + 6O_2 \xrightarrow{\text{yeast}} 6CO_2 + 6H_2O$$
$$+ \text{energy (688 Kcal/mole)}$$

$$\text{sugar} + \text{oxygen} \longrightarrow \text{carbon dioxide}$$
$$+ \text{water} + \text{energy (heat)}$$

This is the process of respiration which is necessary for most organisms and which is used by yeasts when adequate oxygen is available. Yeasts need oxygen to divide and, in a fermenting must (the technical term for juice used in winemaking), a rapid build-up of yeast cells proceeds while there is still dissolved oxygen in the liquid. When the oxygen becomes scarce the respiration and yeast growth slow down and anaerobic fermentation increases. Organisms obtaining energy by respiration are termed 'aerobic' while those which can use the process of fermentation for energy are 'anaerobic'. If oxygen were to be bubbled through a fermenting liquid, the process would change to respiration and no alcohol would be produced.

More about yeasts and fermentation

In the previous section it was mentioned that microscopic, single-celled yeasts were present in the air. In vineyards and cellars, where wine has been made for many years, the concentration of such yeasts is very high. Unfortunately there are present, in the air about us and on the implements we use, many other micro-organisms — such as spoilage yeasts and many bacteria—which may not be beneficial to winemaking. These use sugar, alcohol or other products in the must as sources of energy and leave behind materials which may cause off-flavours or unpleasant odours. One of these is the acetic acid bacteria (*Acetobacter* species) which, in the presence of air, converts alcohol to acetic acid (vinegar).

It is very fortunate for winemakers that wine yeasts do multiply anaerobically and that their products, alcohol and carbon dioxide, inhibit the growth of other micro-organisms. There are two further properties of wine yeasts which are extremely valuable—the first is their ability to tolerate high levels of sulphur dioxide. Sulphur dioxide is a gas which dissolves in water and inhibits the growth of most micro-organisms; products which bear the label 'Contains Preservative' usually contain sulphur dioxide. The use of sulphur dioxide will be discussed later. The second advantageous property is the tolerance of wine yeasts to solutions which are relatively acidic in nature (as low as pH 2.8)—most other micro-organisms do not multiply at such low pHs. Acetic acid bacteria do grow, but need oxygen; lactic acid bacteria, for malolactic fermentation, referred to later, grow at pH 3.2 but not below.

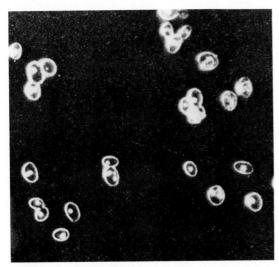

Figure 11.1 Yeast cells

Winemakers must, at all times, observe three rules if they wish to have confidence in the success of their wines:

Rule 1. Always observe the strictest hygiene. Micro-organisms are everywhere, but are especially numerous in the dirt and left-over products of a winery. Clean out all equipment that has been used and sterilize whenever possible. Sterilization will be the subject of later discussion.

Rule 2. Keep air away from fermenting must and wine. Without oxygen most bacteria and fungi will not be able to grow and spoil a wine. There will be fewer problems when the wine is fermenting rapidly, since the evolution of carbon dioxide will keep the air away from the surface of the liquid. Because of this, wine is sometimes fermented in open containers. Once, however, fermentation is complete the wine should be covered and air should be excluded. Many winemakers prefer to keep containers airtight during fermentation, and this is certainly a better policy. If the must has to be held for any length of time before fermentation begins, it should be cooled, sulphured, kept in a sealed container, and the air in the head space replaced by carbon dioxide gas.

Rule 3. The use of sulphur dioxide is recommended at all times. Details of how to use sulphur dioxide will be given later, but it is important that it should be added as soon as the grapes are crushed and that the level be maintained at correct concentrations until the wine is in the bottle. If the concentration is too high it may inhibit the yeast and spoil the quality of the wine. If it is too low, it will not be effective against harmful organisms.

Once the single yeast cell has reached its optimum size it will, under suitable conditions, form buds which break off and begin life as a new yeast cell. This method of multiplication can be very rapid and the aim of the winemaker should be to build up the population of yeast cells in a fermenting must as quickly as possible before harmful micro-organisms can gain a foothold. Sometimes winemakers use wild yeasts, which will already be present in the must, and wait for them to accumulate. However the delay in getting the fermentation underway can cause problems. As we have seen, in areas where wine production has been established for a long time, and the yeast composition in the vineyards is of a desirable type, fine wines can be made without the addition of *cultured* yeasts. Even here the winemaker assists natural fermentation and encourages the growth of wine yeasts (*Saccharomyces* species) by special methods described later under 'Yeast strains and inoculum'.

Cultured yeasts

There are different sorts of yeasts but not all of them are ideally suited to fermentation. The most common yeast is *Saccharomyces cerevisiae* and some of the strains are better for specific purposes than others. Micro-biologists have collected strains and, by cultural techniques, can provide the industry with selections guaranteed to make a satisfactory wine—providing of course that other conditions are favourable. It may be too expensive to use the pure cultures alone to begin fermentation, since an optimum number of yeast cells must be provided; in fact, 10^6 cells per millilitre. Winemakers increase the yeast numbers by making starter cultures as will be seen later.

Wild yeasts

When starting a wine with wild yeasts the initial fermentation may be accomplished by a number of genera, species and strains but, as the alcohol concentration reaches between 4 and 6 per cent, most of these are inhibited and the wine yeast becomes predominant. Unfortunately, even if the wine yeasts take over fermentation at this stage, poor flavours may occur due to other products formed initially by wild yeasts. Rapid methods of pre-fermentation handling, settling of grape must and the correct use of sulphur dioxide will help the wine yeast grow at the expense of the other undesirable types.

Nutrition of yeasts

Yeasts need carbohydrate for growth but, just as we could not live solely on sugar or starch, so also do yeasts need other nutrients in order to multiply. The necessary materials are nitrogen and other minerals, including trace elements, and certain vitamins. Grape juice is particularly good for winemaking because it normally contains all requirements for the healthy growth of yeasts.

Table 11.1 shows the major constituents of grape juice. Not all the components are used for yeast nutrition; some, such as tannin, have important effects on the resultant wine, but little on the yeast. Acids, too, are not directly important as yeast foods, but they do affect their growth. Wine yeasts, as already noted, grow well in acidic solutions, whereas many other organisms are inhibited.

The addition of yeast foods such as di-ammonium phosphate (DAP) is becoming a standard precaution against the formation of off-odours caused by poor yeast nutrition during fermentation of both red and white wines.

Table 11.1 The major constituents of grape juice*

Sugars	Glucose	8-13%
	Fructose	7-12%
Organic acids	Tartaric	0.2-1.0%
	Malic	0.1-0.8%
	Citric	0.01-0.05%
Tannins	Catechol, chlorogenic acid, caffeic acid	0.01-0.10%
Nitrogenous compounds	Amino acids and proteins	0.03-0.17%
Other minerals	Phosphates, sulphates	Traces
B-Group vitamins	Thiamine, riboflavin, pyridoxine, nicotinic acid	Traces
Ascorbic acid		Traces
Volatile aroma constituents		Traces
Colour constituents		Traces

* Amerine and Joslyn, *Table Wines, The Technology of their Production*, University of California Press, 1970

Malo-lactic fermentation

Malo-lactic fermentation is a special process due to lactic acid bacteria which converts malic acid to the weaker-tasting lactic acid. It normally occurs after primary fermentation. Under some conditions, where the wine produced is too sour to the taste, winemakers encourage malo-lactic fermentation. This is particularly the case in cool grape-growing areas. Malo-lactic fermentation may produce desirable flavours in a wine, in addition to reducing acidity, but it can also produce off-flavours. As in the primary alcoholic fermentation, spoilage will depend on the strains of bacteria and yeast present and the conditions under which the fermentation takes place.

Although the lactic acid bacteria can grow under anaerobic conditions and will tolerate high levels of alcohol, they will not multiply under high, free sulphur dioxide concentrations. After the primary fermentation is completed the free sulphur dioxide level is minimal and, if the wine is left under these conditions for any length of time, malo-lactic fermentation may occur. Large quantities of carbon dioxide are not produced but, if the wine is sampled, it may taste 'gassy' or 'tingly' on the tongue. The most crucial factors which affect malo-lactic conversion are temperature, the availability of oxygen from the atmosphere, pH, free sulphur levels and the presence of *lees* —the sediment formed by decomposing yeasts and bacteria. These are discussed in more detail later.

Temperature

The higher the temperature of a fermenting must, the quicker the reaction will be. Under warm conditions, 20-30°C (68-86°F), fermentation will be completed in one to two weeks; under cool conditions, below 20°C (68°F), the time of fermentation is two, three or more weeks. It is not desirable to get fermentation over as soon as possible. White wine is considered to be superior if fermented under lower temperatures: 10-15°C (50-59°F) is considered to be ideal. Red wine can be safely fermented at higher temperatures of between 20-30°C (68-86°F). Because heat is given off during fermentation, very high temperatures can be achieved and these may result in 'stuck', i.e. stopped, fermentation and off-flavours. Wineries, with large quantities of must fermenting rapidly, experience problems of overheating, especially in hot climates, and various methods to cool the wine are used.

Sugars, acids and pH

Winemakers often use terms other than percentage to describe the amount of sugar in a must. These may cause confusion at first, for example, *Degrees Brix* means the percentage of dissolved solids in a liquid, *Degrees Balling* is effectively the same as °Brix. In most wines the majority of the dissolved solids are sugars and the °Brix is the same or slightly more than the percentage of sugar. Another term sometimes used is *Degrees Baumé* which is approximately half the value of °Brix. Generally, if the Brix is known, and an approximate conversion to Baumé is required, it is necessary to divide the Brix by two and add one.

Thus $18°\text{Brix} = \dfrac{18}{2} + 1 = 10°\text{Baumé}$;

$22°\text{Brix} = \dfrac{22}{2} + 1 = 12°\text{Baumé}$.

Sugar can also be estimated by determining

the specific gravity, which is the weight of a given quantity of must divided by the weight of an equal quantity of water. Another term sometimes encountered is Öchsle and this is the weight by which one litre of must is heavier than one litre of water.

Methods of determining the sugar level of must or wine, plus conversions of °Brix, °Baumé and Öchsle, are given in Appendix 1

At harvest, grape juice for wine should ideally contain 6-10 g tartaric acid equivalents per litre (about 0.6-1.0 per cent acid). This level produces a wine with a sharpness which is satisfactory to the palate: higher levels produce wines which are, generally, too acid for most tastes.

Increased acidity is normally accompanied by a reduction of the pH. The correct pH is important for satisfactory growth of yeasts. Optimum pH levels are between 3.1 and 3.5. The methods used for measuring acidity and pH are more complex than those used for determining the levels of sugar, and a certain amount of laboratory equipment is necessary if satisfactory results are to be achieved—see Appendices 1 and 3.

When pH levels are too low 'de-acidification' can be achieved, and the correct levels reached by the addition of amounts of calcium carbonate (food grade), or calcium carbonate-based commercial products like 'Acidex'.

A pH level which is too high, or an acid level too low, may, respectively, be reduced or increased by the addition of organic acids. A mixture of tartaric and citric acid in a ratio of 5:1 is suitable for this purpose. Minor additions of malic acid are also used in finished wines to provide a slightly fruitier flavour.

De-acidification with 'Acidex' or calcium carbonate

Must or wine with excessive acidity can be de-acidified using a double-salt method with 'Acidex'*.

A pre-determined volume of must is de-acidified with an exact amount of 'Acidex' (see Table 11.2) in tank A as shown in Figure 11.2. It is then pumped through a filter into the holding tank B, containing the rest of the must to be treated. It is recommended that the must should be treated prior to fermentation. The manufacturer's instructions should be followed to achieve the best results.

The accompanying Table assumes that 1,000 litres of must are to be treated and the example below shows a calculation to determine necessary amounts of 'Acidex' and must for specific requirements.

---

* 'Acidex', if not readily available, can be obtained from C.H. Boehringer Sohn, 6507 Ingelheim am Rhein, West Germany. 'Acidex' is a commercial preparation of calcium carbonate which may substitute for 'Acidex'.

Table 11.2 Reduction ot acidity by 'Acidex'
Figures reter to a total volume of 1000 litres of wine to be treated

Actual acidity of must before treatment	1.2% Acidex kg	Must litres	1.1% Acidex kg	Must litres	1.0% Acidex kg	Must litres	0.9% Acidex kg	Must litres	0.8% Acidex kg	Must litres	0.7% Acidex kg	Must litres	0.6% Acidex kg	Must litres	0.5% Acidex kg	Must litres
1.0%									1.3	260	2.0	350	2.7	420	3.4	495
1.05%									1.7	290	2.4	380	3.1	450	3.7	520
1.1%							1.3	225	2.0	320	2.7	400	3.4	495	4.0	560
1.15%							1.7	260	2.4	350	3.1	435	3.7	520	4.4	590
1.2%					1.3	190	2.0	290	2.7	380	3.4	480	4.0	550	4.7	620
1.25%					1.7	225	2.4	320	3.1	420	3.7	495	4.4	580	5.1	640
1.3%			1.3	160	2.0	260	2.7	350	3.4	435	4.0	530	4.7	610	5.4	670
1.35%			1.7	190	2.4	290	3.1	380	3.7	465	4.4	550	5.1	630	5.8	680
1.4%	1.3	160	2.0	225	2.7	320	3.1	400	4.0	495	4.7	570	5.4	640	6.1	690
1.45%	1.7	190	2.4	260	3.1	350	3.7	420	4.4	510	5.1	590	5.8	660	6.5	700
1.5%	2.0	225	2.7	290	3.4	380	4.0	435	4.7	520	5.4	600	6.1	670	6.8	710
1.55%	2.4	260	3.1	320	3.7	400	4.4	465	5.1	530	5.8	610	6.5	680	7.1	720
1.6%	2.7	290	3.4	350	4.0	420	4.7	480	5.4	550	6.1	620	6.8	690	7.4	730
1.65%	3.1	320	3.7	380	4.4	450	5.1	495	5.8	570	6.5	630	7.1	700	7.8	740
1.7%	3.4	350	4.0	400	4.7	465	5.4	510	6.1	580	6.8	640	7.4	710	8.1	750
1.75%	3.7	380	4.4	420	5.1	480	5.8	520	6.5	590	7.1	650	7.8	720		
1.8%	4.0	400	4.7	435	5.4	495	6.1	540	6.8	600	7.4	660	8.1	720		
1.85%	4.4	420	5.1	450	5.8	510	6.5	550	7.1	610	7.8	670	8.4	730		
1.9%	4.7	435	5.4	465	6.1	520	6.8	570	7.4	620	8.1	680	8.7	730		
1.95%	5.1	450	5.8	480	6.5	530	7.1	580	7.8	630	8.4	690				
2.0%	5.4	465	6.1	495	6.8	540	7.4	590	8.1	640	8.7	700				

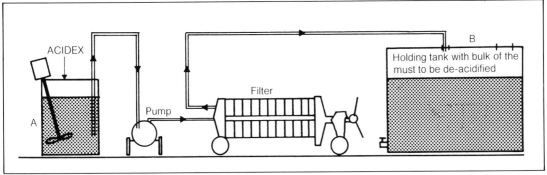

Figure 11.2 De-acidification of musts and wines using the double-salt technique with 'Acidex'

Example

Must to be de-acidified=910 litres at 1.75 per cent acidity. Final desired acidity=0.8 per cent. From the Table it can be seen that 6.5 kg of Acidex and 590 litres of must are needed (tank A) for 1.75 per cent actual acidity to be de-acidified to 0.8 per cent.

The correction for the required volume, 910 litres, is as follows:

$$\frac{590 \times 910}{1,000} = 537 \text{ litres which requires}$$

$$\frac{6.5 \times 910}{1,000} = 5.92 \text{ kg 'Acidex'.}$$

Acidity is often expressed as grams per litre; 1%=10g/l

Extract

Extract, also referred to as 'sugar-free extract', consists of a diverse group of minerals and non-volatile substances dissolved in wine. Higher extract levels help give 'body' to wines and indicate a greater ripeness and quality of the grapes. Low extract levels indicate the use of unripe grapes, or wine dilution, and are naturally less desirable. In general, the level of extract in table wines varies from 17 to 30 grams per litre of dry wine. The higher the amount of extract, the greater the 'mouth-filling' quality of the wine.

Sulphiting and other precautions

The value of adding sulphur dioxide to musts to reduce the activity of wild yeasts and undesirable micro-organisms, while allowing the wine yeasts to grow satisfactorily, has already been mentioned. There is, however, another reason for adding sulphur. Once a plant tissue is cut, or crushed, enzymatic browning occurs and this can produce undesirable brown colours in wine and certain off-flavours. Additionally, oxygen from the air will cause other chemical reactions in crushed fruits which can have deleterious effects on wine, and the use of sulphur dioxide will reduce these oxidative and enzymatic reactions. It is very important, therefore, to reduce the time needed to begin fermentation and to introduce sulphur to the musts as soon as possible. To do this, most winemakers add sodium or potassium metabisulphite, which is easy to store and use. If 200 mg of this is added to one litre of must it produces the equivalent of a concentration of 100 ppm (parts per million) sulphur dioxide. Sulphurous acid is also sometimes used for the same purpose. This is used as a 5 per cent solution which is added at 2 ml per litre of must to give a sulphur dioxide concentration of 100 ppm. For large-scale production, sulphurous acid may be prepared by bubbling compressed sulphur dioxide from cylinders through water. Details may be obtained from suppliers.

Sulphur dioxide at 50-80 ppm may be added at the time when grapes are crushed or during the preparation of must. A simple way of doing this is to dissolve the crystals in a small quantity of juice which should, subsequently, be thoroughly mixed and dispersed in the must or mixed with the crushed grapes. Not all the sulphur dioxide is available for the purposes described above and about three-quarters is immediately bound to other chemicals and made inactive; this proportion increases as fermentation proceeds. Chemical tests are available to determine both the total and free levels of sulphur dioxide, see Appendix 4. There is no point in adding sulphur dioxide during fermentation, as this would lead to the formation of excess aldehydes and other undesirable off-flavours and would have, anyway, no beneficial effects at this stage.

When a must is vigorously fermenting.

carbon dioxide gas is produced and this effectively prevents oxygen gaining access to the liquid. If, before or after fermentation, an excessive amount of oxygen is in contact with the must or wine it will initiate some of the undesirable changes which have been previously described. While sulphur dioxide will reduce these changes, a further precaution is to introduce carbon dioxide gas over the crushed fruit and into the containers to reduce the risk of aeration. It is also important that musts should be kept at a low temperature and that wines should be handled rapidly and as little as possible.

Further information about the preparation of musts is given in the chapters relating to specific types of wines—see Chapters 12 and 13.

Chaptalisation

'Chaptalisation', or the sugaring of grape must, is an acceptable practice in cool-climate wine regions. In cold seasons only low levels of natural sugars are produced which results in wines which lack fullness of body and flavour, and have an insufficiently high alcohol level to be stable and harmonious.

The addition of sugar to must or fermenting wine is carried out in different ways. The direct addition of sugar to actively-fermenting must is practised by some winemakers, others first dissolve the amount of sugar required in a small amount of must and then add this, prior to fermentation, to the settled must. The latter method is preferred since additions of undissolved sugar to an already fermenting or fermented wine can result in 'stuck' or stopped fermentation which can be difficult to restart. Losses can also occur because part of the sugar may remain undissolved and settle out with the lees.

Some dilution of wine flavour and bouquet occurs with all but the smallest of sugar additions—although the final wine appears superior to that made from less-ripened grapes not sweetened in any way which will tend to lack body. Grape concentrate can also be used for sweetening unripe grape must; this method is more expensive and is usually used only in areas where cane sugar cannot legally be added to must or wines (e.g. Australia, California and.South Africa).

The term 'enrichment' is sometimes preferred to chaptalisation since this more-general term covers all sweetening materials—e.g. sugars, grape concentrates, etc.

Containers for fermentation

The essential feature about containers used for fermentation is that they must be made of a material which does not react chemically with the wine. Wine must never be placed in contact with any material containing copper or iron. The only suitable metal is stainless steel and, even here, care must be taken in selecting a suitable grade. For example, Grade 316 is satisfactory while Grade 304 is less so. Stainless steel is being used increasingly in winemaking and many modern wineries are almost entirely stocked with stainless steel equipment. Concrete is also suitable for fermentation vats, and wineries occasionally use concrete water-storage tanks and paint the inside with two to three coatings of a polyurethane paint, or blow-torch applied layers of wax, before adding the musts. Fibre-glass lined metal tanks can also be used and some wineries are experimenting with plastic tanks. Finally, oak casks have been traditionally used for fermenting and ageing wines but, their cost ensures they are mostly used only for better wines.

Whatever materials are used, it is important that they be properly cleaned before use. Stainless steel should be washed with a hot solution of 1 per cent potassium carbonate and then thoroughly washed with water. Glass, concrete and fibre-glass can be washed with a 1 per cent solution of citric or sulphurous acid.

Oak casks need special treatment. Casks which have held astringent or highly-flavoured products, or are mouldy or damaged, should not be used. Casks which have previously held red wine contain high amounts of tartrate and colouring deposits and are unsuitable for white wines. Tartrates themselves can be removed by soaking with a solution of 3-5% sulphuric acid (H_2SO_4) for 24 hours, this is followed by a thorough washing with water. Steam from steam-generating plants can be used to clean the inside of casks.

Casks should never be allowed to dry out and, if a cask is to be stored for a long time, it is wise to leave it full of a 1 per cent solution of potassium carbonate. Before use, new casks should be soaked for seven days with a solution of sodium metabisulphite or sulphurous acid at the previously-mentioned concentrations. With old casks a 24-hour period will be sufficient. These soakings should then be followed by a further 24 hours in a 1 per cent solution of

(a) (b)

(c) (d)

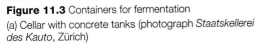

Figure 11.3 Containers for fermentation

(a) Cellar with concrete tanks (photograph *Staatskellerei des Kauto*, Zürich)

(b) Steel tanks indoors pre-filtration with Kieselgur—diatomaceous earth (photograph *Wädensville Institute* Switzeriand)

(c) outdoor steel tanks (photograph *Montana Wines*)

(d) Indoor steel tanks (photograph *Montana Wines*)

potassium carbonate. The casks are then rinsed with water until no smells are left.

When pumping must into a fermentation container it is important that it should not be filled to the brim: as fermentation proceeds a froth is formed which could overflow the container. If the must contains grape skins which float to the surface—as with red wines —the container should not be more than two-thirds to three-quarters full, and its contents should be stirred two or three times a day whilst fermentation is in progress.

It has already been stated that it is desirable to ferment in closed containers in order to keep oxygen from making contact with the must. Often the container will be sealed off by a pressure valve or fermentation trap.

Yeast strains and inoculum

In France, Germany and many other European winemaking countries, the fermentation process is, as we have seen, sometimes begun by a natural infection with yeasts from vineyards (mixed cultures). The most commonly used method is to crush sound berries from the best part of the vineyard about a week before the others, and to allow the resulting must, sulphited to 100 ppm, to ferment in warm conditions. This 'starter' culture is then added to the sulphited must and produces rapid fermentation which eliminates the effect of undesirable micro-organisms and assures the quality of the final product. The use of these 'mixed' natural cultures is only possible where the micro-organisms in the district are predom-inantly of the right type, (e.g. Saccharromyces spp).

Research has shown that even though there is a wide range of yeast strains, only a few are responsible for quality wine production; therefore, to limit the effect of undesirable yeasts, the use of cultured yeasts is recommended. Pure cultures will not only promote complete fermentation and tolerate high levels of sugar, alcohol and sulphur dioxide—they will also ensure constant character and quality, especially under cool conditions. There are, in addition, specialised yeast cultures such as low-volatile acid-producing strains useful in dealing with botrytised and otherwise damaged grapes; others are known to produce high ester levels for aromatic wines, some are more tolerant of low temperatures or very high sugar concentrations.

The wine-maker will probably wish to experiment with different, commercially-available yeast cultures before deciding which is the most suitable for a particular wine. Although two different strains of yeast may produce slightly differing types of wines, these differences will not be as great as those which derive from varieties of grape, levels of maturity and methods of fermentation. Thus, it is not possible to produce good wine simply by the choice of a good yeast.

Yeast cultures are available from the trade, either dry or contained in liquid or agar.

A culture of yeast is a concentrated collection of yeast cells specially prepared by the manufacturers. These may be added directly to the must to begin fermentation, but it is more usual, especially if large quantities are required, to increase their number by growing them under ideal conditions before adding them to the must. This means that fermentation will begin almost straight away and there will be less opportunity for harmful micro-organisms to build up.

A sterilized 'starter-medium' may be prepared by boiling 350 ml of juice for 15 minutes, or steaming it in a pressure cooker or autoclave for 5 minutes. This is the sterilized starter medium. Containers for this should preferably be Pyrex glass and should be carefully sterilized. After sterilizing, make sure the liquid or the containers are not left open to the air otherwise they will be recontaminated. Then add the yeast culture to the sterilized starter medium contained in a 500 ml conical flask. Do not add the yeast until the temperature of the medium has dropped to 25-30°C (77-86°F). The flask should be sealed with a bung of cotton wool. Maintain the mix at a temperature between 20-30°C (68-86°F) to encourage rapid fermentation. When the medium is fermenting rapidly the process can be repeated — adding this prepared yeast culture at the rate of 5 per cent to further, starter mediums which have been sulphited to 100 ppm, but not boiled. The process continues, at the same rates, until the total volume is between 2 and 5 per cent of the must that is to be used for wine. At 20-30°C (68-86°F) it will take about three days for each starter medium to ferment adequately and thus the winemaker can estimate approximately how long he should allow to provide sufficient starters for his estimated volume of wine. Dried yeast cultures, used at about 20g/100L (0.9g per imp. gal. or 0.75g per US gal.) concentrations, are first re-hydrated in a small amount of warm (40°C) solution of must and water and will be ready for use, e.g. activated, within a few hours.

Once a container of wine has been started, a sample of its must can be used for starting another batch of wine; but care should be taken that wine spoilage or disorders are not spread in this way and only sound must should be used.

It is important not to allow the temperature of the fermenting must to rise too high. Many modern wineries have refrigerated and pressurized, stainless-steel, fermenting tanks which accurately control the rate of fermentation. As we have seen, white wines, particularly, benefit by being fermented at low temperatures. It is certainly best to choose cool conditions for fermenting since, as has already been seen, off-flavours can be produced by the fermentation reaction which raises the temperatures of large quantities of must to levels which are too high. Small quantities of wine in cool conditions rarely suffer any problems in this respect. If the level of sugar drops by 1°Brix a day, the rate of fermentation is satisfactory and temperature control is unnecessary.

The end of fermentation is indicated by a drop in the rate of carbon dioxide release which, when it starts to decline, quickly falls to zero. Musts, however, sometimes get 'stuck' which means that fermentation stops before all the sugar has been used up or the alcohol has risen to a sufficiently high level: when this occurs it is important that it should be re-started as soon as possible. This can be done by reinoculation with fermenting must from another batch; it might be helped by slight aeration and by the adjustment of temperature and pH if these are outside the desirable range.

Almost all wine yeasts will become inactivated by the time the alcohol levels reach 14-15 per cent, even if there is still sugar in the must. The amount of sugar left after fermentation has finished determines the residual sweetness of the wine. A sweet wine will have 4 per cent or more sugar; a semi-sweet wine about 2 per cent, and a dry wine below 1 per cent. There are chemical methods of determining residual sugars in wines (see Appendix 1), but some simpler kitsets, for example Clinotest, are available which make the job much simpler.

If, after fermentation, all the natural sugar is used up, the wine can be sweetened by adding further sugar or 'sweet reserve'. In the EC sugar may not be used for sweetening wine. Sweet reserve will be described in detail in the next chapter but, basically, it is unfermented grape juice which is added to the wine after fermentation. It must be remembered that adding sugar or sweet reserve to a wine may cause additional fermentation and, if this occurs in the bottle, it could lead to a cloudy product. Thus the yeast cells will need to be inactivated and removed from wine before bottling. The most satisfactory way to do this is to use sterile filtration and other methods described later in this chapter.

The amount of alcohol that can be produced in a must is slightly less than half the percentage of available sugar, see Appendix 5. However, as has been already indicated, wine yeasts become inactivated after alcohol levels reach 14 per cent, and a sugar content of over 28 per cent will not raise the alcohol above this level. Alcohol levels can be measured by methods described in Appendix 5.

Malo-lactic fermentation

The use of malo-lactic, sometimes called 'secondary' fermentation, is common — especially with red wines — in most cool-climate areas of Europe. It reduces the effective acidity, gives stability, and allows prolonged bottle-ageing. However, it must be completed before the wine is placed in bottles, otherwise secondary fermentation can produce serious faults in bouquet and flavour.

Malo-lactic fermentation converts malic acid into lactic acid and also releases small amounts of carbon dioxide, ethanol and acetic acid. The overall consequence is to reduce the effective acidity of the wine. These changes are brought about by species of bacteria belonging to the genera *Leuconostoc, Pediococcus* and *Lacto-bacillus*, which are present in most wineries and normally enter the wines during handling.

Malo-lactic cultures, like yeasts, are available in dry or liquid form for addition to wines. Usually they are incorporated towards the end, or after primary fermentation is complete.

Conditions favourable to malo-lactic fermentation are as follows:

1. pH. If the pH is outside the desirable range of 3.2 to 3.4, winemakers may decide that a small amount of chemical acidification or de-acidification may first be necessary.

2. Temperature. If the temperature is below 15°C (59°F), bacterial activity is inhibited and, in cool climates, warming of wines may be practised. On the other hand temperatures close to 40°C (104°F), although satisfactory for bacterial growth, may result in a loss of wine character.

3. Nutrients. If the wine is racked and filtered immediately after the completion of primary fermentation, the essential nutrients may have been removed. To overcome this problem, filtering is delayed until after the malo-lactic fermentation is complete. Remember, however, that long contact with lees (sediment) can cause off-flavours.

4. Oxygen. It is important that no sulphur dioxide additions are made to the wine after primary fermentation and, in addition, it is desirable to have slight aeration to encourage growth of the bacteria. A single racking may provide the required aeration. Excessive amounts of carbon dioxide or nitrogen gas should never be introduced into the wine in which secondary fermentation is to be encouraged. Containers in which malo-lactic fermentation is to be induced should be kept full of wine.

5. Bacteria. A tank or cask in which malo-lactic fermentation is underway may be used as a starter culture. This can be added at a level of about 5 per cent to the appropriate wine. As we have seen it is now possible to buy cultures of suitable bacteria which can be progressively reared, i.e. activated, and added to the wine in a manner similar to that described for yeasts. The wine should be tasted regularly and tested for pH in order, effectively, to monitor the changes which are occurring. Additionally, the changes in acids can be gauged by the use of paper chromatography as described in Appendix 7.

Failure to provide for any of these conditions may either delay, slow down or prevent malolactic fermentation from occurring. If conditions, however, are ideal, the process should be completed within a few weeks after primary fermentation.

The clarification and stabilisation of wines

After fermentation is completed most yeasts will have stopped growing and sunk to the bottom where, with other sediment, they are known as lees. The wine at this stage will have no free sulphur dioxide left and, providing malo-lactic fermentation is not required, it is important to re-adjust this to an appropriate level. This adjustment will facilitate and speed up sedimentation and clarification, and will prevent oxidation and reduce the danger of bacterial contamination. Generally speaking, 50 to 80 ppm needs to be added to provide a free level of about 20 ppm. As a general rule, if a high amount is used before fermentation, a high amount will be needed for the adjustment.

Although a dry wine, at a normal pH of 3.0 to 3.6, will require a free level of 20 to 30 ppm, wines with 2 per cent sugar need 30 to 40 ppm and, if sugars are 5 per cent, the wine will need 50 ppm of free sulphur dioxide. If the wine is dry but the pH is 3.6 or more, a free sulphur dioxide level of at least 30 ppm is required for adequate protection. Once the winemaker has decided on the required level of free sulphur dioxide, he can determine the amount of metabisulphite needed by adding, say 30, 50 and 80 mg to a litre of each of three samples of wine. These should be left overnight and the level of free sulphur dioxide determined, see Appendix 4, to give an idea of the amount required to achieve the desired levels. The level of sulphur dioxide in young wines should be checked at 14-day intervals and adjusted as needed. The level in different wines will vary.

Although much of the solid material in the wine will be incorporated in the lees, there is still usually a fine haze which needs to be eliminated. This is mainly due to proteins which, by themselves or in combination with phenols, form minute particles which remain in suspension. Clarification means the removal of these particles by racking, fining and filtering. Following clarification the wine should be stabilised to preserve its appearance and soundness after bottling.

It is worth noting here that, although the exclusion of oxygen in containers is recommended, controlled oxidation can be beneficial to enable the rapid growth of yeast in early fermentation, to enhance bouquet and flavour development, and increase colour stability in red wines, whilst premature ageing is avoided. Such oxidation may occur during crushing, pressing and racking; of course it must not be overdone.

Racking

Racking means pumping or syphoning off the wine from the lees and this should be done as soon as possible after fermentation, otherwise off-flavours may develop due to the decomposition of yeast cells. It may need to be done two to four times before the wine becomes clear. Sulphur dioxide levels will need to be adjusted to the required level over the racking period.

Fining and cold stabilisation

The number of rackings may be considerably reduced by the use of finings. These materials cause the suspended proteins, which give the wine its cloudy appearance or haze, to aggregate together and sink to the bottom where they are more easily separated from the clear liquid by racking and filtering. Rapid removal of proteins prevents their subsequent decomposition and the release of off-flavours.

There are many different fining agents and a wide range of opinions on how to use them. The effectiveness of fining is not equal to the quantity used, but to how and when the winemaker uses it. Two such agents are bentonite and gelatine.

Bentonite

Bentonite is the most common fining agent and is readily available, cheap and easy to use. A wine with a light protein haze can be cleared by using 0.5g per litre, but heavier hazes need from 1 to 4 g per litre of wine. Laboratory tests on 1 litre lots of wines should be carried out to determine the correct amount of bentonite that is required. The best way to use bentonite is first to mix the required amount with a small quantity of hot water. The ratio of 1g: 20ml of water gives easy mixing and a good dispersion of the bentonite in the wine — necessary factors for quick and effective fining. The wine will need to be left for one to two days to achieve adequate clarification. Small additions of bentonite to a white must prior to fermentation (0.5 to 1.0 g per litre,) improves protein stability and early clarification of young wine and reduces the risk of spoilage due to the decomposition of lees.

Gelatine, Tannin-gelatine

Gelatine added at 0.06 g per litre will act as a protein fining agent and has the additional effect of reducing tannin levels. Where bitterness, e.g. excessive tannin, is a problem, gelatine is used alone, but otherwise tannin will need to be added with the gelatine for satisfactory clarification, as with most white wines. It is important to determine the correct ratio of tannin to gelatine for each wine and, to do this, 0.06 g of gelatine is added to each of 10 one-litre test samples of wine; to the first sample add 0.2 g of tannic acid and to the second 0.4 g and so on until the tenth sample, which will contain 2 g of tannic acid. Mix thoroughly and allow to settle for a few hours. The correct proportion for the particular wine will be determined by selecting the sample whose ratio gives the best clearing effect. Some winemakers, however, use a combination of both methods — tannin-gelatine during the settling of white must before fermentation, followed by bentonite after fermentation. The use of gelatin fining in young, light-coloured red wines will result in colour loss, therefore it is not recommended.

Cold stabilisation

Sometimes a crust, consisting of salts of tartaric acid, is found on the inside of wine bottles: although these crusts are not harmful to the wine, they do reduce consumer appeal. It is possible to prevent this crusting by using a cold stabilisation technique which brings out the crystals prior to bottling and which, at the same time, ensures that no further crystalisation takes place during storage. The wine is cooled to -4°C (25°F) for four days, at which stage the solubility of the salts is reduced and they precipitate out. The wine is then filtered while still cold.

After fining, the wine can be racked into storage containers for maturation, or directly into bottles if maturation prior to bottling is not required. The more diligent winemaker, as has already been indicated, will check the free sulphur dioxide level every time a wine is moved and, if necessary, adjust it to a suitable level (between 15 and 45ppm free SO_2). It is also common now to filter the wine before bottling, to make doubly sure that no haze or sediment enters the bottle.

Filtering

Various filters are available commercially, most of which simply remove particles suspended in the wine. Where, however, wine has residual sugar or sweet reserve is added, it is necessary to use sterile filtering. Such filters are so fine that almost all micro-organisms are removed and the wine is, therefore, virtually sterile. It must be emphasised, especially after sterile filtering, that bottles and corks must also be sterile, and the wine must be transferred immediately from the filter to the bottle with minimum exposure to air. Sterilisation can be rendered more effective by purchasing bottles in sterile packs which are opened just prior to bottling. The use of sterile membrane filters—attached to a bottling line—is highly recommended. Corks can also be bought in sterile packs: if not sterilised they should be treated by soaking overnight in a 1 to 2 per cent solution of sulphur dioxide. Soaking for a longer period than this is dangerous and will damage the cork. As an additional precaution, during bottling, the air space above the wine in the bottle can be filled with nitrogen from a cylinder.

It is often noted that the filtering of a wine creates a rather flat taste which takes about six weeks to disappear. If the asbestos pad of the filter is first rinsed with a 1 per cent solution of citric acid, or steam sterilised, the effect is not so pronounced. Nowdays asbestos-free filter pads are commonly used—these do not incur this problem.

Preservatives

Sorbic acid at 200 ppm acts as a stabilising agent and prevents general spoilage by bacteria and the re-occurrence of fermentation in wines which contain residual sugar. It is added to the wine, in the form of potassium sorbate dissolved in water, prior to filtration and bottling. Sweeter wines will require a higher level than dry wines.

It should be remembered that the use of sorbate as a wine preservative can cause off-flavours—especially with sweet wines—and

Figure 11.4 Plate filter
Wines are filtered to improve their clarity and stability.
Wine filters form an essential part of winery equipment

its use is not recommended for quality wines which require bottle-ageing.

Sulphur dioxide, at concentrations of 200 to 450 ppm, provides a similar protection although, at these high levels, the sulphur's pungent odour can be unpleasant. Modern winemakers, as a consequence, prefer to use other methods of stabilisation so as to keep the levels of sulphur at a minimum. Winemakers should check local regulations to find concentration limits for the use of sulphur dioxide and other materials.

Diethylpyrocarbonate, DEPC, (marketed as 'Baycovin'), combined with sterile filtering, is sometimes used for the stabilisation of both sweet and dry wines. The wine is first filtered through a sterile, asbestos filter under inert atmosphere—nitrogen or carbon dioxide—into a sealed tank. The DEPC, mixed with a small amount of wine, is then added and thoroughly dispersed in the wine to be bottled. Levels of 80 to 100 ppm for dry and 200 ppm for sweet wines are recommended. During this process it is essential that the wine should not come into contact with air.

Depending on temperature, DEPC is effective for a limited period of time—at 20 °C (68°F) for 48 hours — and, after this time, it breaks down to harmless by-products (carbon dioxide and ethyl alcohol), which are already present, in far greater quantities, in the wine. The wine should be bottled before the DEPC has reached its break-down stage.

During the bottling operation some winemakers, as an extra precaution, follow the addition of DEPC with a further sterile filtering, this time using membrane filters. A second filtering is a valuable precaution, especially when bottling sweet wines.

The sterility of bottling-line components—filter-head lines, corks, bottles, etc—should be checked regularly during bottling to ensure the long-term stability of wines.

It is important to note that the general trend in winemaking and wine legislation is towards a reduction of the use of preservatives. This is made possible by more rapid methods of handling wine and the use of modern filtration systems.

Cellar handling of damaged and mouldy grapes

In some years, a portion of the harvest may contain mouldy or otherwise damaged grapes. To produce a sound wine from such grapes, the winemaker will have to employ cellar-handling techniques which are in some ways different to those used in normal vintages. Common causes of grape damage include; hail, bird damage, berry splitting, harvester damage, and poor weather at harvest. Typical secondary infections of damaged fruit will involve various fungi (moulds) such as *Aspergillus* or *Botrytis*. Further damage will be caused by a rapid spread of acetic-acid bacteria, often spread by the 'wine fly', i.e. *Drosophila*.

Cellar-handling techniques employed for handling damaged fruit aim particularly to prevent oxidation, the build-up of acetic acid, and the development of mouldy odours and off-flavours. The following steps will assist in the production of sound, fault-free wines from a damaged crop.

• The rapid separation of grapes from contact with air, which is the source of atmospheric oxygen and secondary infections by moulds and bacteria.
• The use of inert gas (carbon dioxide or nitrogen) blanketing in sealed containers during grape transport.
• Early and sufficiently high sulphiting of the grape must. Levels of 100-150 ppm of total SO_2 are considered sufficient to prevent the growth of spoilage micro-organisms such as wild yeasts and bacteria.
• Rapid crushing and light pressing to achieve a fast separation of must from the grape skin. Once again, the use of inert gas blanketing will prevent oxidation. Prolonged skin contact promotes the development of mouldy odours and off-flavours and thus should be avoided.
• The early reduction of temperature of the

grape must to below 5°C This will be especially important with grapes harvested during warm conditions.

- Cold settling of solids from the musts is beneficial and will be greatly improved by the use of finings. The high levels of free proteins found in mouldy grapes can be reduced to tolerable levels by bentonite or gelatine fining. The most appropriate levels should be determined by a laboratory trial (already described in this chapter under the heading, *'The clarification and stabilisation of wines'*). When a very strong mould odour is observed, the must can be treated with PVPP polyvinylpyrrolidone or carbon finings, though it must be observed that high levels of such finings will diminish the colour, bouquet and flavour characteristics of the finished wine. For the use of PVPP or carbon finings follow the manufacturer's instructions and always determine the quantities of PVPP to be used by trial on 1-litre lots.

- The removal of mouldy off-flavours and odours is more easily achieved and more complete if done prior to fermentation.

- In large, well-equipped wine cellars cold settling is usually replaced by centrifugation and pasteurisation (rapid heating for not more than five minutes at 60°C, 140°F) of musts from damaged harvests. This is faster and equally effective in removing free proteins and inactivating oxidative enzymes which are naturally present.

- Once clear and stable, the must from mould-infected grapes should be fermented as soon as possible. Inoculations with 3 to 4 per cent pure yeast starter, actively fermenting, is important. The use of special low volatile acid-producing yeasts can be helpful. The addition of further bentonite at 1g per litre, can be beneficial; this will be especially important with musts cleared to very low, solid levels.

- Monitoring of acetic acid levels at all stages of settling and fermentation is recommended. A build-up of acetic acid is a sure indicator of activity by spoilage micro-organisms. Such activity, when observed, will require pasteurisation of the fermenting wine and the re-introduction of an active, healthy, yeast starter. The use of sulphur dioxide during fermentation should be avoided at all costs, as it would only achieve the development of an excessive amount of aldehyde and sulphur-based odours and off-flavours in the wine.

- The removal of sulphur-based off-odour compounds such as mercaptans and hydrogen sulphide can be achieved by the addition of small amounts of copper sulphate at 0.1-0.4g per 100 litres of fermenting wine. Copper sulphate should be dissolved in a small quantity of water first and is much better added during, rather than after, fermentation. The use of this fining — which precipitates insoluble sulphide compounds—is sometimes restricted, the local regulations controlling additives must be consulted.

- Excessively low fermentation temperatures below 15°C are not recommended as these have been shown to promote the development of acetic acid in wines made from mouldy grapes. Experimentation with different yeast strains for producing wines from mouldy grapes in different locations would be desirable.

- After fermentation, early clarification and protein stabilisation will be required. The potassium levels can be abnormally high in wines made from mouldy grapes and, where this was not adjusted prior to fermentation, it must be done immediately after the fermentation is complete. The addition of meta-tartaric acid, followed by cold stabilisation and filtering, should reduce potassium to tolerable levels and achieve acid stability.

- Early bottling of wines produced from damaged grapes is desirable. The wine is adjusted to 30-40 ppm free sulphur dioxide and an inert gas is used to protect the colour of the wine.

The basic steps in the making of red and white wines, described in the next chapters, are illustrated graphically in Figure 11.4.

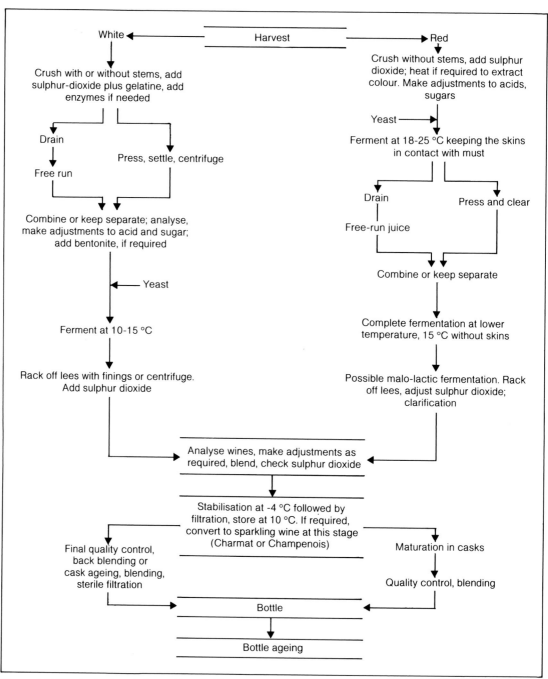

Figure 11.5 The basic steps in the making of white and red wine

CHAPTER 12

THE PRODUCTION OF WHITE AND SPARKLING WINES

There are several distinct processes in the production of white wine which will be considered under separate headings. These are harvesting, crushing, pressing, fermentation, clarification and maturation. Some processes have already been mentioned and this chapter will look, in more detail, at those aspects which are specific to the production of white wine. Only an outline of the methods used for making sparkling wines will be given, as the production of such wines involves additional and sometimes rather elaborate techniques.

Harvesting

Determining maturity

As grapes ripen, concentrations of sugar, flavour and aroma increase and levels of acid decline. The winemaker must decide when his grapes are at their optimum stage for the production of the desired wine. Generally white grapes should reach at least 16°Brix, while acidity levels should have fallen to between 0.6 and 1.0 per cent tartaric acid equivalents. In favourable seasons these levels will be attained in most vineyards; in cold years one or both of these standards may not be reached and the grower and winemaker will need to take corrective measures.

The following adverse seasons pose specific problems for the grape grower and winemaker.

Cold season. A cold season where the sugar levels remain low and acid levels are high causes many problems. The grower will realise that, no matter how long he leaves the grapes on the vine, they will not completely ripen before the advent of wet cool weather in autumn and that problems of damage and disease may arise. He must therefore at some point decide whether the additional ripening that may occur is more than offset by the dangers of disease. The resultant wine will not be top quality but, by the careful addition of sugar and a reduction of acidity, an acceptable and even a good wine can be made.

Warm, humid and wet season. A warm humid and wet season can cause just as many problems as a cold season. Under these conditions the temperature is theoretically adequate to promote sugar production and to reduce acidity. However, vegetative growth may still persist late in the season, causing the humidity around the grape to increase, and disease to become a problem. If the wet weather persists the grower may pick early to avoid an excessive loss of crop. The result will be a must with low sugar, high acid and diluted character.

Hot and dry season. A hot and dry season is not common in cool-climate wine districts. Sugar production is adequate, and may even be high enough to produce a naturally sweet wine, but acid levels are often too low. Cool-climate wine grapes grown under such conditions seldom develop full flavour and bouquet and rarely produce fine-quality wine. Nevertheless, a wine of good, even standard can be produced. It is worth noting that some of the disadvantages of hot, dry years can be overcome by early picking and rapid separation of skins; by gelatine fining, low temperature settling and fermentation, and by early bottling.

Most problems in harvesting will be caused by one or other of the above factors although the last one—a hot dry climate—is not so relevant to this book. The best white wines seem to be produced in the areas of the world which tend to be on the cool side and which are subject to the first set of seasonal conditions. Germany and northern France are the prime examples. A cool, long growing season, where grapes ripen in the autumn, allows for the maximum development of the many known and unknown components of a fine wine. Temperature variation in the final stages of ripening is thought to be an important factor.

Before looking at methods of harvesting, it is now necessary to consider special types of wine which are dependent on certain growing conditions. For the purpose of easy reference these wines will be grouped under their German nomenclature. The best quality wines of this type are produced in cool climates. In warm areas, late picking brings other complications such as low acid, high pH and coarse 'raisin' like flavours.

Spätlese (late harvest). Sometimes in a good year a grower will leave a proportion of his crop to be late-picked. These grapes develop more sugar and more flavour components, and the subsequent wines are a little sweet and reach fine quality. The extra cost of the wines reflects the extra risk the grower takes by leaving his grapes for an additional period—about two weeks—on the vine. In parts of the Rhine, Riesling for spätlese must be at least 75 degrees Öchsle (18 °Brix), although it should be remembered that sugar level alone is not important and that high fruit flavour concentration must be achieved.

Late-picked grapes often become infected with *Botrytis* ('Noble Rot'). If conditions are cool and moderately dry, the fungus will live on the surface of the fruit and extract moisture from the grape without causing it to collapse. The grapes shrivel and the sugar, acid, flavour and aroma constituents become highly concentrated. The following wines will have an increasing degree of *Botrytis* infection.

Auslese. If, from a late crop of grapes, the ripest bunches showing 'Noble Rot' are specially selected, the resultant wine is of even finer quality than spätlese and can be called an *auslese* wine. In Germany, the sugar levels will be at least 90 degrees Öchsle (21.4 °Brix) and the wine will often have a pronounced *Botrytis* character.

Beerenauslese. Sometimes a further selection can be made of the best berries from chosen bunches to make a *beerenauslese*. In Germany, the sugar levels are a minimum of 120 degrees Öchsle (28 °Brix).

Trockenbeerenauslese. A wine made from selected, shrivelled grapes from the best bunches is considered to represent the absolute peak in German wines and is called a *trockenbeerenauslese*. Tremendous prices are paid for these wines which are much sought after but available only in very limited amounts. Sugar levels must be 150 degrees Öchsle (35.4 °Brix), or over.

It is worth noting that the prime growers of Sauternes in France also rely on the Noble Rot mould to increase sugar concentration and to produce sweet, high-quality wine.

Eiswein (Ice Wine). Very occasionally a frost will affect the berries of late grapes. When this occurs ice crystals form within the berry which concentrate the remaining juice. To produce this exceptional wine the winemaker must pick the berries while still frozen and press out the juice before the ice crystals thaw.

In practice the German grower not only uses the level of sugar as a guide, but also considers the winemaking methods to be used and the flavour development of the grape. Each district stipulates the sugar levels which, in that area, will normally produce the required quality— *spätlese, auslese*, etc. In areas outside Germany the sugar levels which have been cited may be a less reliable guide to prime quality, and certainly they would need to be tested over a number of years to determine the level which produces the best wine.

Experienced wine growers can decide when to harvest by tasting their grapes. In districts where large areas are under cultivation and winemakers buy their grapes from contract growers, wineries may train a panel of tasters to gauge grape quality and ripeness in the vineyard. The value of juice sample-tasting is being increasingly recognised in many parts of the world. Payment for the grapes may vary according to the assessed value of the grape for quality wine. This is backed by the use of quantitative methods.

The use of refractometers and hydrometers to measure sugar, and the methods used to determine acidity and pH are described in Appendices 1, 2 and 3.

As maturity approaches the grower should make tests once a week. The sampling of grapes is important and a minimum of 200 berries is required. Bunches should be chosen from different levels of the vine—on both the shaded and sunny sides. Three berries should be selected from each bunch, one from the top, one from the middle and one from the base. Great

Mechanical harvester operating in a large vineyard

care must be taken that the sampling is truly representative and that all parts of the vineyard have been covered.

Picking the grapes

The small winemaker will pick by hand but mechanical harvesters are now available which considerably reduce the picking costs on large vineyards. These not only pick the grapes but they can also be combined with a field crusher which will crush them while still in the vineyard. Mechanical harvesting is not recommended for wine styles made from over-ripe or botrytised grapes. Lack of selection, excessive damage and yield loss would occur with this type of harvest.

Picking should be done as carefully as possible so that the berries are not damaged. The juice from broken berries is subject to enzymatic oxidation and spoilage, which occurs readily in warm temperatures, and causes off-flavours and unsatisfactory colour in wines. Oxidation is highest at 30°C (86°F) and considerably lower at temperatures below 10°C (50°F) and, for this reason, picking is best done on cool days or in the early morning. Boxes of grapes must be kept in the shade and may be covered in the vineyard. The time between picking and crushing should be as short as possible to prevent oxidation and the entry of unwanted yeasts and bacteria which cause spoilage of the wine. Part of the sulphiting can be done in the vineyard and consists of sprinkling a measured amount of sodium metabisulphite over the grapes in the field, or on the trucks if the grapes are to be carried considerable distances from vineyards to cellar.

Crushing and de-stalking

Crushing simply means the breaking of the skins of the grapes so that juice can escape; thus fermentation in red wines is possible and the pressing of whites is more effective. Stalks are normally removed from the grapes before crushing begins as they contain a high level of phenols which, if they were crushed with the grape, could make the resulting wine astringent and bitter. The removal of stalks is usually done by machines which will often both de-stalk and crush the grapes. Some winemakers, however, crush the grapes without removing stalks and, if done gently, it does not interfere with quality; it is used to improve pre-draining and pressing efficiency. It also gives a higher yield of must with a lighter degree of pressing. An important consideration is the design of the de-stemmer. Some older models are known to strip stalks to such an extent that the release of bitter-tasting substances is inevitable. The winemaker will need to experiment with his own grapes to see whether crushing with stalks is feasible or not.

Spray residues may be poisonous to yeasts and man: they will often interfere with fermentation and cause off-flavours in the wine. Before harvest, therefore, it is essential to note the waiting period recommended for each spray and to allow this time to elapse between spraying and harvest.

After crushing, a certain amount of free-run juice collects at the bottom of holding or receiving containers and this may be drawn off and fermented separately. About 30 to 40 per cent of the total volume of juice may separate as free-run juice which, generally, is considered to produce a superior quality of wine. Draining is a standard procedure in many winemaking operations and special drainers are used to assist the process. Essentially they have a columnar stainless steel mesh cylinder

into which the juice moves and from which it can be run off. A more sophisticated drainer is shown in Figure 12.1.

Normally, after crushing, the grapes are immediately transferred to the press, but sometimes, as additional constituents may be drawn from the skins, it is considered valuable to leave the crushed grapes in a container for a limited period of time, say 24 to 48 hours. In such an event it is important that the grapes be adequately sulphited, the containers are closed, and the air about the grapes replaced by carbon dioxide. If the process is continued too long, especially under warm temperatures, undesirable materials such as excessive phenols will be extracted. This can be limited by adding gelatine in a proportion of 0.1 g/litre to the crushed fruit.

As a general rule skin contact is recommended for the aromatic group of varieties, e.g., Muscats, Gewürztraminer, Müller-Thurgau. Varieties which prove difficult to press, such as Sylvaner, will also benefit from skin contact, especially if this is combined with the addition of pectolytic enzymes to the crushed fruit.

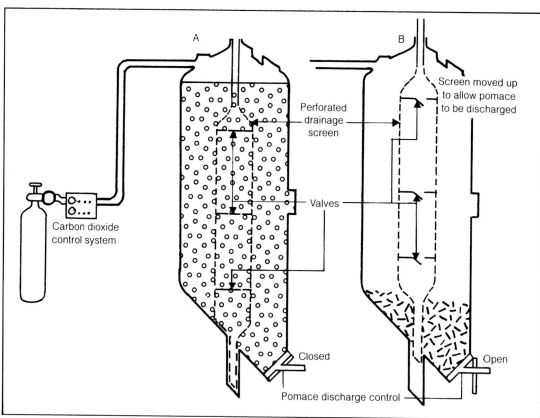

Figure 12.1 Modern juice separator with pressurised drainer tank

Enzymes

A number of pectolytic enzymes are now commercially available which, when added to crushed grapes, increase the volume of free-run juice and assist the pressing operation. These enzymes promote the separation and breakdown of cell walls and thus help release the juice in the cells. Amounts to be added are given on the label and will not be repeated here.

The use of enzymes is increasing and is generally favoured by the winemaker, who thereby obtains larger wine volumes. As with all techniques, its use needs to be carefully monitored to ensure that, concommitant with volume increase, quality does not decline.

Pressing

There are several types of presses commercially available which will separate the juice or must from the skins. It is important to understand that the degree of pressing will affect the subsequent quality of the wine. Free-run juice —perhaps 30 per cent of the total—and a light pressing to remove a further 40 per cent will generally contain the finest quality of juice for winemaking. A winemaker may decide to ferment this portion separately and market it as a premium-grade wine. A further one or more pressings might then be adopted and the juice used for standard-grade wines. Pressings should never be so severe that seeds are ruptured, since these contain unacceptably high levels of phenols which will be released into the must. If oxidation is a problem, a blanket of carbon dioxide gas can be used during the pressing to cover the pulp.

Settling of must

If not added during the crushing operation, sulphur dioxide should now be incorporated in the juice. After pressing, must will often benefit by being allowed to settle for 24 hours in cool conditions—preferably close to 0°C (32°F), after which the partially-cleared juice is pumped from the sediment. A limited amount of bentonite fining at 1.5 g/litre can be added to help the sediment to settle. Although some winemakers maintain that the use of bentonite at this stage will adversely affect the bouquet, flavour, and body of the wine, this method is quite widely used in Germany and local winemakers maintain that it produces a cleaner bouquet. Centrifuges can sometimes be used to clear the must and can considerably speed up the operation. The settled or centrifuged wine clears faster after fermentation and gives smaller amounts of more compact lees.

Settling must be undertaken in sealed conditions, cooled to below 10°C (50°F), and the air above the must should be replaced with

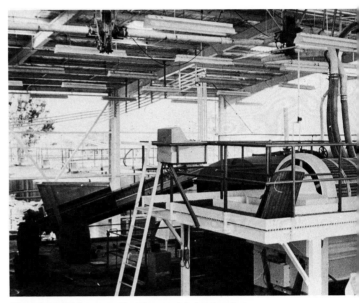

Figure 12.2 Vintage cellar. The careful planning that goes into winery design is shown in this modern cellar equipped with the latest winemaking machinery. Fully automatic reclining de-juicer and Vaslin press in the foreground, temperature-controlled stainless steel fermentors in the background, and overhead must lines are seen above (Photograph: *B Seppett and Sons*).

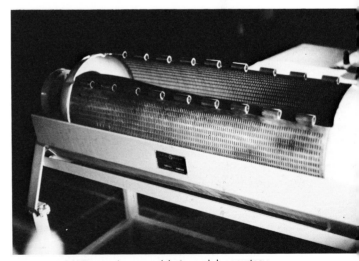

Figure 12.3 Willmes wine press (photograph by courtesy *Geisenheim Wine Research Institute*)

carbon dioxide. It should be noted that excessive clarification at this stage may make it difficult to complete fermentation and may, as a consequence, result in wines that lack body and varietal character. This specifically applies to less-ripe vintages.

Blending and additions to the must

In the section on determining maturity it was said that certain modifications will be needed if the acid and sugar levels in grapes are unsuitable. If the acid levels are high they will not only make the resulting wine too sour, they will also reduce the pH level and, if this falls below 2.9, fermentation will be slow— or may not occur at all. It is at this stage that calcium carbonate or commercial products like 'Acidex' are used to raise the pH to between 3.0 and 3.6 (see Table 11.2). If low acidity and high pH are a problem, a 5:1 mixture of tartaric and citric acid, or tartaric acid alone, can now be added.

The best way to correct poor balance, additives notwithstanding, is by blending. Simply put, musts containing excess acidity are blended with others containing too little; those with high sugar are blended with those with low sugar. For this reason it is often helpful to grow several varieties of grapes which ripen at about the same time. Settled musts can be stored for up to ten days for blending purposes, provided they contain adequate sulphur dioxide, are sealed, covered with carbon dioxide, and provided the temperatures are kept at a minimum, preferably below 5°C (41°F). Alternatively, they may be blended after fermentation to alleviate problems with storage. As a general rule major additions and blending is best done before the fermentation starts.

Fermentation

Although fermentation has already been discussed in some detail, it could be useful to re-emphasise that low temperatures, preferably 10-15°C (50-59°F), are especially important for white wines since more of the volatile components of grape juice are retained at these temperatures, and the production of compounds — such as aldehydes, acetic acids, higher alcohols and esters—are reduced. White wine has a more delicate flavour than red and is more dependent on the retention of desirable components and the elimination of undesirable ones. When using low temperatures it is most important that the yeasts chosen for inoculation are adapted to these temperatures,

otherwise the rate of fermentation will be too slow.

More and more wineries are now changing to stainless steel or metal pressure-fermentation tanks which are sealed, and which usually have some method of reducing temperature. As pressure increases in the container, the rate of fermentation is reduced. Thus, by modifying the pressure at which carbon dioxide is released, the winemaker can control the rate of fermentation, as well as the final sweetness of the wine.

Once fermentation is proceeding, the free sulphur dioxide is quickly lost and there is no point in adding more. A must with a suitable yeast strain which ferments at a steady rate— about one degree Brix reduction per day— cannot easily be damaged and, in this condition, it is least susceptible to undesirable microorganisms.

Storage and ageing of white wine

After clarification and stabilisation, white wines can be stored in inert containers, matured in oak casks, or bottled straight away. Whichever is used, it is important to keep the wine at low, approximately 15°C (59°F), and even temperatures, to prevent contact with air, and to maintain sulphur dioxide at a suitable free-sulphur-dioxide level. In order to prevent air contact, the volume above the wine should be flushed out with an inert gas. The use of carbon dioxide during storage and bottling will give a freshness and crispness to the taste. If it is wished to avoid this, nitrogen can be used.

White wines may be fermented and stored for a limited period in oak casks to add character to certain varieties such as Chardonay or Pinot blanc. If the wine remains too long in the wood, it loses its bouquet and gains an excessive oak flavour. Wine in casks should be tasted at weekly intervals and should be removed when experience tells the winemaker that it has obtained its optimum benefit. Some wine will be lost by evaporation and barrels, if stored upright, should be topped up with fresh wine. It is especially important that the sulphur dioxide levels be maintained and that the casks are sealed.

The average time in barrels for a white wine is 6-10 months. The newer the barrel, the less time is required.

All operations after clarification are best undertaken under cool, dry conditions. These conditions should also be sought for the storage of wines in containers, casks and bottles. Bottles

The bottling operation is done by modern automated machinery (photograph courtesy Montana wines Ltd.)

should be stored on their sides to keep the cork moist and to prevent the access of air to wine.

White wines do not normally benefit from long storage and many are sold and drunk soon after bottling. Nevertheless, some ageing is desirable, especially for acidic wines, and better winemakers will favour this practice. Light, fruity wines will reach their peak of perfection soon and will probably begin to deteriorate after two to five years. Specialist wines like *ausleses* and *beerenausleses* will last for longer periods of time and improve with several years' storage.

Generally, more complex wines like Chardonnay, Pinot blanc, Riesling, Sauvignon blanc, Sémillon, are stored for longer periods than simpler wines such as Chasselas, Müller-Thurgau, Sylvaner and the Muscats. It is not possible to make too many rules about the length of storage and the winemaker must experiment with his own product if he wishes it to mature before it is sold.

The use of sweet reserve

Except in hot climates, or under conditions of late harvest with *Botrytis* infection, sugars are nearly all converted to alcohol. Quite often a winemaker wishes to make a sweet or semi-sweet wine and to do this he can use one of four methods. Firstly, he can add additional sugar to the must in order to create a level at which the yeast cannot completely ferment. The remaining sugar will impart sweetness to the wine. This is not generally used, since the resulting wine often lacks balance. Secondly, by chilling followed by sterile filtering to eliminate micro-organisms and prevent further fermentation, the winemaker can stop fermentation before it

has finished. The third way, is to use a sweet reserve. Sweet reserve is an unfermented must which is added to finished wines. It is normally added just before bottling and the mixture is sterile-filtered directly into the bottles. After pressing, a proportion of the juice is separated from the bulk and stored for one or two days under carbon dioxide to settle the solid particles—bentonite may help this operation. The clear juice is separated off, sulphited to a free level of 100 ppm and filter-sterilized into sealed and chilled containers filled with carbon dioxide. If DEPC is used, the level of sulphur dioxide may be reduced to 40 ppm. The sweet reserve can also be preserved by 'fortification' with alcohol at 16 per cent (not allowed in the EC). It is usually filtered and stored at low temperatures and under inert atmosphere. The fourth method of making naturally sweet wines from botrytised grapes is described in the next section .

The production of naturally-sweet, botrytised wine

In many respects the production of botrytised wines will follow the principles outlined for the handling of damaged and mouldy grapes discussed in the previous chapter. The prevention of acetification, oxidation, and the problems with high potassium and protein stability will be the same. In addition, however, the special aspects pertaining to the *botrytis* character, the acid/sweetness balance and astringency, resulting from the prolonged skin contact required, must be considered. Some of the principles involved in the making of sweet, botrytised wines will be summarized below, though the experience of the winemaker in handling these wines will be important if quality, auslese, or sauternes style wine is to be made in commercial quantities.

Much of the success with botrytised wines will depend on the suitability of the region for developing *Botrytis* mould in bunches of ripe grapes. The most appropriate climate is one where the grapes mature under cool conditions with little rain. Dews and fogs caused by cool nights following warm days assist the growth of the fungus — *Botrytis cinerea*. Lack of warm, wet conditions prevents the fungus growing too quickly and reduces the attack by other undesirable fungi and bacteria. The berry should shrivel, but not split.

Bunch and even berry selection may be undertaken to achieve the desired character such as spätlese, auslese, etc. (See 'Harvesting', earlier in this chapter).

The extraction of the *Botrytis* character will depend on a suitably long holding period of the harvested fruit on the skins. Up to a week can be required to produce a honeyed character although, unfortunately, astringency from phenols and the danger of browning due to the action of enzymes will also increase. Gelatine is used to reduce astringency at 20-30g per 100 litres at crushing—and inert gas blanketing and suitable sulphiting should avoid uncontrolled fermentation and browning. Because botrytised musts have greater sulphur binding capacity, additions of up to 150 ppm sulphur dioxide may be necessary to achieve 30-40 ppm free.

The exact amount of time required for holding the grapes on the skins is determined by a regular analysis of the sugar level in free-run juice; when the level does not change further the grapes are pressed. This may be modified by tasting the juice to assess the amount of *Botrytis* character—the desired amount will also influence the decision to press.

Levels of acid, pH, and high levels of potassium are best adjusted prior to fermentation. Because of the sweet nature of the wine, more acid is required to balance the sugar— levels of 10-12 g/litre are desirable; botrytised musts can have up to five times the levels of potassium found in some grapes.

Fermentation in steel tanks or barrels should proceed at approximately 15°C and can be stopped when a desirable flavour and sweetness is found in the wine. Botrytised German style wines usually have lower alcohol levels and although the presence of *Botrytis* acts as a yeast

inhibitor, the winemaker must take precautions that the wine will be stable. To stop the fermentation process, the wine is chilled to 0°C, clarified by centrifugation and sterile filtered. Racking and earth filtration are not usually adequate for such wines. Where rapid chilling and centrifugation is not possible, fermentation will not be stopped and the level of sweetness in the final wine may need to be supplemented. This can be done by adding sweet reserve.

For botrytised, sweet wines, the final free sulphur dioxide levels are higher than normal —up to 50 ppm. The complex bouquet and flavour characteristics of botrytised sweet wines will not be obvious for some time after bottling. Cellar ageing for one to three years will be necessary to achieve the character demanded of these highly-priced wines. Further bottle-maturation before consumption is recommended.

Due to the nature of their production, the French Sauternes or Barsac wines reach considerably greater alcohol levels than German wines. The winemaker in Sauternes will rely on the high alcohol level of 13-15 per cent, and successions of sulphur additions during a number rackings in barrels, gradually to stabilise his wine. This is made possible by the greater ripeness and botrytis infections which occur with Sémillon and Sauvignon blanc grapes grown in these two districts of Bordeaux.

The production of sparkling wine

The base for making any sparkling wine can be either a red, white or rose wine, stabilised and cleared to a stage when it would be normally ready for bottling. There are three basic methods for producing sparkling wine: *Carbonation, Charmat* and *Champenoise*. The basic principles for each are described below.

Carbonation

When ready for bottling, the base wine is usually chilled and carbon dioxide is injected into the wine under pressure during the bottling operation. It is the simplest and cheapest way of producing a sparkling wine, but on release of the cork the sparkling nature quickly disappears as it does with any carbonated drink.

Charmat method

In this method a finished base wine is placed into large pressurised tanks and a small preparation of sweetening and yeast starter is added to recommence alcoholic fermentation. Since the carbon dioxide cannot escape into the atmosphere it remains dissolved in the wine. When completed the now sparkling wine is sterile-filtered and bottled under pressure. The majority of quality sparkling wines in all countries are made in this manner.

Champenoise

This method is used to produce the best quality sparkling wine and is only adopted when the finest base wine is available, as in the Champagne district of France. The steps in this process are described, although it should be noted that individual winemakers will use slight modifications to suit their own style, conditions and grapes.

1. Blending (Coupage). Considerable care is taken to blend base wines. In the Champagne district, wines from both red and white grapes are used; the juice of the red grapes is separated from the skins before colour extraction in special presses. Base wines are blended to create specific styles. The Blanc du blanc style is produced from white grapes only. Vintage Champagne is made from wines produced in a single year.

2. Sweetening and the addition of yeast (Liqueur de Tirage). Cane sugar is used to sweeten the base wine and allow fermentation to recommence. Yeast culture is also added at this stage. Approximately half of the sugar will be converted to carbon dioxide. The addition of 25 g of sugar per litre of wine is recommended and a small amount of citric acid is sometimes included to assist the conversion of sucrose to glucose and fructose.

3. Bottle fermentation. The wine is placed into special 'Champagne' bottles, and capped. The very slow, second fermentation takes between three and five months to complete. Bottles are placed on their side at 5-8°C (42°F) in long stacks and left to ferment and to mature for a period of some years. At regular intervals the bottles are shaken to keep the lees from settling.

4. Clearing (Remuage). Bottles are placed on specially designed shaking-tables called *Pupitres*. On these the bottle position is changed from almost horizontal to upside-down in three or four weeks. Each day, every bottle is very gently shaken and turned slightly to force the sediment into the neck of the bottle; when inverted this sediment rests firmly against the cap. This process is now commonly mechanised

5. Recorking (Degorgement). The deposited lees are removed from the wine either by releasing the cap while rapidly moving the bottle from an upside-down position to upright (the ancient method), or by first freezing the contents in the neck of the bottle and then, with a special tool, removing the uppermost layer of the frozen wine in which the impurities are trapped. The second method is invariably used as it requires less skill and considerably less wine is lost.

6. Sweetening and final corking (Dosage). The wine lost during recorking is replaced with a *liqueur d'expédition* usually consisting of a brandy-wine base, with honey or cane sugar. The final sweetness is determined by the amount of the sweetening added (Brut 1.5 per cent, Dry-Sec 4 per cent, Demi-Sec 8 per cent and Doux 12 per cent by volume). A proper Champagne cork is then inserted (plastic ones or beer caps are usually used for fermentation), the cork is wired and bottles are then placed in the cellar for bottle-ageing. It may take four to five years before bottle-fermented, sparkling wine reaches the market.

One special champagne style is made by ageing on the lees for prolonged periods of five, ten or more years prior to clearing. This gives added complexity, richness in yeast character and greater flavour concentration.

CHAPTER 13

THE PRODUCTION OF RED AND ROSÉ WINES

The main difference between red or rosé wines and white wines is that the must is fermented after crushing with the skins still present, and pressing takes place later. It has already been mentioned that the juice from red grapes is, with most varieties, almost colourless and it is the skins which contain the colouring material, plus tannins and other flavour components which give a red wine its distinctive character. During fermentation, the yeast and alcohol draws out the pigments from the skins and the wine gradually assumes a red colour. The degree of colouring depends on the variety of grape; some grapes will never produce more than a light pink or rosé wine, while others will produce a deep red colour. The length of time for fermentation will determine the degree of coloration and it is not essential that the must should be completely fermented in order to achieve the desired result. Rosé wines are often fermented for just one to two days and then pressed, after which fermentation is continued without skins. In cooler climates with Pinot noir grapes, or other varieties with a typically low colouring, the ratio of juice to skins can be improved by drawing off some of the free-run juice after it has been crushed and fermented separately as a white wine. The remaining red wine will, as a result, have greater depth of colour and tannin content.

Harvesting

Oxidation is less of a problem with red grapes, although harvesting techniques are still very important. However, the lesser problem of oxidation notwithstanding, it is still advisable that winemakers adopt the methods which have already been recommended for harvesting white grapes—especially when it seems probable that insufficient colouring will appear.

Mechanical harvesting of Pinot noir and other fine-quality red wines is not generally recommended. Professor Peynaud, in his textbook on winemaking in Bordeaux,* gives detailed reasons for this conclusion. Even in new-world regions nearly all fine red wines are being produced from hand-harvested grapes. This, of course, is not to say that sound, red table wine cannot be made from mechanically-harvested grapes, but that the same degree of control for specialised wines cannot be achieved.

Crushing

In many parts of the world the crushed red grapes are fermented without sulphur dioxide, but the addition of 50 ppm at the time of crushing is recommended. Stalks are usually removed before crushing, but the skins and the seeds remain in the fermentation vats. Retention of some or all of the stalks is practised for some red wine styles (see *Maceration carbonique*).

Fermentation

It is quite common for red wines to be fermented in open containers. There are three main reasons for this. Firstly, oxidation is not such a problem; secondly, the skins float on the surface of the fermenting liquid and, in sealed containers, are difficult to mix into the

---

* Peynaud, E. 1984. *Knowing and Making Wine*— translated from French by A. Spencer, Wiley, New York.

Figure 13.1 Vintage in Bordeaux
Grapes are harvested into bins on the backs of workers and then placed Into containers

Figure 13.2
After harvesting, the grapes, in tractor-drawn containers are taken to the cellar for crushing.

Figure 13.3 Fermentation of red wine—wine being pumped to mix skins and liquid (photograph *Montana Wines Ltd*)

must. Thirdly, contamination and off-flavours are more easily masked by the stronger character of red wines. Nevertheless, towards the end of fermentation, or at the beginning if it is slow in starting, possibilities of contamination are present and the use of tanks with lids is recommended whenever this is possible.

The problem of the 'cap of skins' has been mentioned and if this is not broken a layer

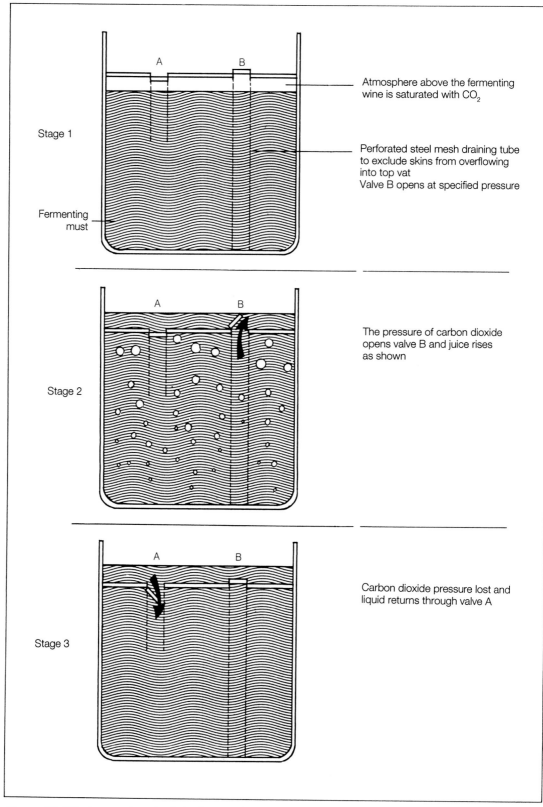

Stage 1

Fermenting must

Atmosphere above the fermenting wine is saturated with CO_2

Perforated steel mesh draining tube to exclude skins from overflowing into top vat
Valve B opens at specified pressure

Stage 2

The pressure of carbon dioxide opens valve B and juice rises as shown

Stage 3

Carbon dioxide pressure lost and liquid returns through valve A

Figure 13.4 Bordeaux system of red wine fermentation

of carbon dioxide gas may separate the liquid from the skins and thus reduce colour extraction. The skins may also trap air and acetic acid bacteria, and other micro-organisms may cause spoilage of the wines. The wine can be either stirred regularly by hand, called 'plunging', or by pumping—say every five to seven hours —or other devices may be employed to mix the layers.

Some, like the system shown in Figure 13.4, use the pressure of CO_2 to bring wine over the caps — after completion pomace may be removed using a drainer such as shown in Figure 12.1.

For the production of bulk red wine in southern Europe, more complex systems are used where wine and pomace are removed every 3-4 days and newly-crushed grapes are added, resulting in continuous wine production. Some quality red wine producers use roto-fermenters which are horizontal tanks rotating in pre-set automatic cycles. The rotation together with internal screens in the tanks are claimed to improve colour and tannin extraction.

When sufficient colour has been extracted for a rosé wine, or for a red wine which might become too tannic, the wine is transferred from the fermentation vat to a press and the skins and other large particles are removed. The fermenting must is then pumped back into sealed containers until fermentation is com-plete. If the wine is to be totally fermented on the skins, as is common in most cool-climate districts, it is important to press immediately fermentation has been completed, otherwise, and especially in open vats, deterioration can occur. Traditional methods of making Bordeaux claret and Red Burgundy allow for maceration (e.g. soaking) of skins with fermented wine after sugar has been used up. Submerging the floating cap of the skins and sealing the wine surface from air is considered essential if such methods are used.

Temperatures used in fermentation are usually higher than with white wines; higher temperatures tend to extract more colour but, if they are too high, excessive amounts of tannin and polyphenols are released which dominate the varietal flavour. High temperatures also favour oxidation and give the wine a brownish colour and a 'cooked flavour'. If low colour is a problem it is better to change the ratio of juice to skins in the manner already described, rather than to increase the temperature. Temperatures of 25-30°C (77-86°F) are probably ideal for red wine fermentation. Lower temperatures of 10-15°C (50-59°F) are more suitable if the wine is pressed off the skins and fermentation is to be continued and completed in closed containers.

Higher fermentation temperatures of up to 35°C are known to occur in the production of the traditional, full-flavoured style of Pinot noir from Burgundy and a few other districts. Such temperatures occur only for short periods during the peak of fermentation activity. The average, during the 12- to 17-day fermentation period of red Burgundy, is close to the figures given above.

Maceration carbonique (whole bunch fermentation)

This technique for the fermentation of red grapes is used principally in the Beaujolais district, where light-bodied, soft red wines are produced. It also plays, in part, a role in the production of Burgundies in the Côte d'Or where a third to a half of the total volume can be so fermented. In many Burgundian cellars, the partly de-stemmed or whole bunches are allowed maceration carbonique fermentation during the first part of the process. Well before all the sugar has been used up, plunging or pigeage begins and fermentation is completed hot, helped by the gradual release of sugar from an increasing amount of crushed berries. In many areas there is an increasing demand for such soft red wines and, for such a style, maceration carbonique with Pinot noir or Gamay noir is ideal.

For the true Beaujolais style, on the other hand, all or part of the crop is fermented, without the grapes being crushed or de-stemmed, in sealed containers. Sulphiting of the grapes and the use of inert gas is important to prevent the excessive build-up of volatile (acetic) acid. Cap plunging is not required and the must is pressed just prior to, or on reaching, dryness.

Compared with the traditional fermenting of crushed, de-stemmed grapes, whole bunch fermentation produces wine of a lighter colour, softer, and with a more aromatic bouquet and a fresh, soft flavour.

Lower fermentation temperatures of 15-20°C (59-68°F) are common, higher alcohol levels are often produced but volatile components tend to be more of a problem, especially if damaged berries are used. Less tannin is removed from the skin, but this is compensated for by the additional extraction from the stalks.

If poorly-ripened grapes with less lignified stalks are used, excess tannin can be a problem. The softness is partly due to a higher breakdown of malic acid by yeasts during the three to five week-long primary fermentation.

Pre-fermentation maceration

A recent development in red wine production is the use of a pre-fermentation maceration of chilled crushed grapes which are held in closed vats for up to 10 days before yeast addition. The crushed must, which is highly sulphited to 150 ppm SO_2, is said to produce wine with greater and more stable colour, more primary fruit flavours and higher total tannins compared with more traditional methods. One such approach, called the Accand method, follows the pre-fermentation maceration with a short hot fermentation and early pressing and is gaining favour with producers of Pinot noir in Burgundy and elsewhere. The use of healthy, ripe grapes chilled to 0-2°C appears to be essential. Overcropped, unripe and mouldy grapes are not recommended for this technique. Not all winemakers favour the method and the high sulphur dioxide concentrations are still somewhat controversial.

Heat treatment of red must (thermovinification)

Heat treatment is another method which is sometimes used for extracting colour. After de-stalking and crushing, the must is heated to 60°C (140°F) and held at this temperature for up to three hours, after which it is cooled to 10-12°C (50-54°F) and 50 ppm of sulphur dioxide is added. Sulphur dioxide should not be added before heating. The must is then fermented with skins — or is pressed and fermented without skins, in a manner similar to that described for white wine. Another method requires heat exchangers and consists of 'flash heating' up to 80°C (176°F) for one to two minutes.

Wines produced by such methods are different from traditionally-made red wines. They need less ageing and are softer and lighter in body; they have less astringency, and sometimes better colour. They should not be compared with traditionally-made wines but regarded as a different style and judged accordingly. Some red varieties, such as the Pinot group, are more suited to thermovinification than others.

Figure 13.5 Barrel maturation cellar (photograph *Montana Wines Ltd*)

CHAPTER 14
SMALL-SCALE WINEMAKING

Descriptions of winemaking, to this point, have been concerned with methods which are adaptable to the commercial production of wine on any scale—small, medium or large. In research establishments throughout the world it is becoming increasingly common to test new varieties and techniques on a very small scale, and special methods have been evolved to ensure that this can be successfully accomplished. Whilst these new methods have been developed, primarily, to meet the needs of research workers who require to make routine tests on new varieties, and for commercial winemakers who wish to modify existing techniques or to test new varieties or blends, they are also suitable for the home winemaker.

The methods described are based on personal experience by one of us (D.F.S.) in France, Germany and South Africa, and on techniques developed at the Geisenheim Institute by Professor H. Becker.*

In small-scale winemaking, because the quantities are so much smaller, there is, at all stages, more surface area in relation to the volume that is exposed to the air. The use of sulphur dioxide, cooling and inert gas, therefore, has a more signficant role in small-scale winemaking than in large-scale production .

Containers

Stainless steel, glass or plastic containers are used for volumes of 4-100 litres (1-22 gallons).

*Becker, H. and Kerridge G. H.. 1972. Methods of small-scale winemaking for research purposes in both hot and cool regions. *J. Aust. Instit. Agric. Sci.* 38: 3-6

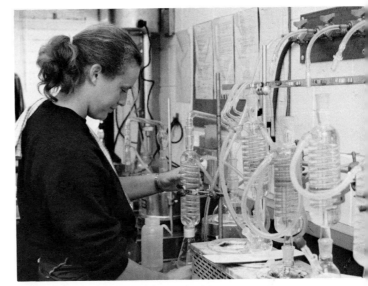

Figure 14.1 Chemical analysis in a wine laboratory

Glass is preferable however, being relatively cheap, free from taint—a problem with some plastics—and enabling the whole process to be observed at all stages.

Procedures for white wines

Crushing and pressing

Grapes should be in good condition, clean and undamaged. They are cooled to 1-2°C (34-35°F) for 24 hours which helps to prevent oxidation and limits spoilage by micro-organisms before and during crushing and pressing: while crushing is in progress 80-100 ppm of sulphur dioxide is added to the grapes.

Figure 14.2 Glass fermentation vessels. Glass fermentors of up to 400 litres are used in the Geisenheim Institute, Germany, for small-scale wines of improved clones with traditional and new grape varieties

Crushing and pressing can be done in a cool room and operations should be performed as quickly as possible while the grapes are still cold. The containers receiving the crushed must should be filled with carbon dioxide just prior to crushing.

The settling of pulp and treatment prior to fermentation

After pressing, the must is placed in sealed containers and the air above flushed out with carbon dioxide. It is then stored for 24 hours at 1-2°C (34-35°F) and the clear juice is siphoned off the pulp which has settled overnight. The addition of bentonite fining, 1-1.5 g/l, after settling to help produce a clear wine and to absorb off-flavoured compounds and protein is often used. The must will now be relatively free from micro-organisms and will quickly begin to ferment once a starter culture is added. The first inoculation of a starter culture can be made into heat-sterilized grape juice, but subsequent transfers to build up the culture should use only unheated juice sulphited to 80 ppm. Starter culture is added at a ratio of one to three per cent.

Fermentation and subsequent handling

Fermentation is best undertaken at a constant temperature of between 10-15°C (50-59°F) and takes two to three weeks. If fermentation stops too soon, the yeast should be stirred to re-activate it. A check on the progress of fermentation can be made by testing sugar levels, initially with a hydrometer then, as the level of sugar falls, with a 'Clinotest' kit. Fermentation should continue until the residual sugar falls between 0.1 and 0.3 per cent. For varietal evaluation in small-scale winemaking it is important to rack the wine off the lees as soon as possible to prevent the onset of malo-lactic fermentation. Some

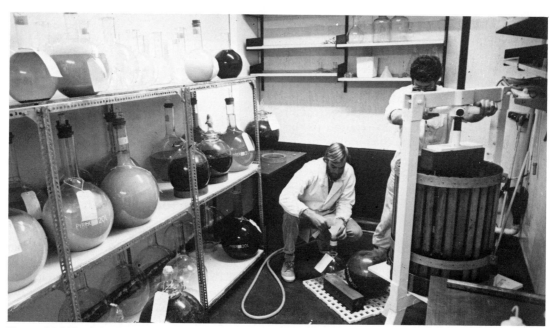

Figure 14.3 Hand pressing a batch of grapes in a small winemaking laboratory

winemakers even rack before the end of fermentation, say about one per cent sugar, to avoid any chance of an off-flavour coming from the decomposing lees. After fermentation and the first racking, bentonite can be used and the second racking is combined with filtration. Sulphur dioxide is added at 100 ppm and maintained subsequently at a free level of 30-40 ppm. The wine is placed in a clean container, the air above the wine replaced with nitrogen gas and the wine is stored for three months at 10°C (50°F) to allow stabilisation. The sulphur dioxide is adjusted as required to a free level of 30-40 ppm. The wine is then sterile-filtered directly into bottles, using pressure from nitrogen gas, and then corked.

Procedures for red wines

The method for red wines is the same as for whites, except that grapes are fermented on the skins at 18-22°C (64-72°F) for a sufficient time to remove colour. Settling and bentonite are not used prior to fermentation. Caps are plunged below the surface five times a day and the whole operation is carried out in sealed containers with fermentation traps. After the colour extraction has been accomplished the must is pressed and the juice transferred to clean containers and fermentation continued at 10-15°C (50-59°F). Further handling is identical to that of white wines.

Other methods of handling red wines

Heat treatment, as described in the previous chapter, is quite easy to adapt to small-scale winemaking. Other methods for extracting colour have also been suggested.

Mashing. Grapes are macerated prior to fermenting on the skins. The colour is quickly extracted and can be up to 100 per cent stronger. Pressing can thus take place earlier than normal with an equivalent or better colouring of the must.

Storage on skins. Crushed grapes are sulphited to 100 ppm, placed in sealed containers under carbon dioxide and stored at 18-20°C (64-68°F) without fermentation, for between 12 and 14 weeks. They are then pressed and fermented without skins.

Use of wood chips

Oak casks are very expensive. Winemakers have therefore experimented with the use of chips or shavings to impart an oak flavour to the wine. One method which, from personal experience, has been found to be quite successful, is to add the chips to the must whilst it is fermenting on the skins in a proportion of 2-3 g/litre, and to press and process it in the normal way. The wine definitely seems to be superior in style to those which have been bottled without coming into contact with oak.

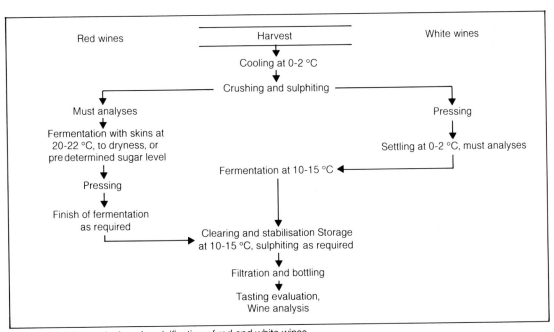

Figure 14.4 Stages in the microvinification of red and white wines

CHAPTER 15

THE STORAGE, AGEING AND SENSORY EVALUATION OF WINES

A wine must be kept for nearly the whole span of man's life in the dark, subterranean solitude of a cellar to become great, fulfilling all its promises, tempering its violence and realising its charm, completing its velvety texture and finally becoming a generous, complex wine whose strength appears tender, which, in its old age will have the purity of an open flower—thanks to a flower that is long since dead . . .

G. Roupnel

Despite the eloquence and charm of the above statement, it should be stressed that not all wines will benefit from lengthy ageing; in fact most of the modern, light white wines retain their freshness only for a short time, and appear best when young. The majority of mass-produced wines belong to this category, though there is no wine which does not benefit from a few weeks' rest after bottling. Modern winemaking practices all over the world aim at the consumption of young wines—freshness being their major asset.

Wine storage in large volume, inert containers

Following fermentation, stabilisation and clarification, young wines are often stored in large steel, concrete or fibreglass-lined, metal containers until bottling. These containers are inert and in themselves add nothing to the wine.

During such storage the container is sealed and the wine is kept constantly at low temperatures, 10-15°C (50-59°F), and a blanket of carbon dioxide or nitrogen gas is used to prevent contact with air and potential spoilage. Bottling capacity and marketing factors rather than winemaking considerations determine the length of storage, since few changes occur in such wine for periods of weeks or even months. Maintenance of an adequate level of free sulphur dioxide (30-40 ppm), coupled with the low temperature and carbon dioxide or nitrogen above the wine, should be the standard practice where this type of storage is used. Where further ageing and maturation are desired, oak casks or bottle-ageing are required.

The maturation of wine in casks

Maturation and ageing of wine in oak casks requires greater attention than the more simple bulk-wine storage in inert containers, since the oak itself adds distinctive bouquets and flavours, and allows a limited exchange of gases between the wine and the cellar atmosphere.

These bouquets and flavours, plus the slow ageing with air exchange, are the specific reasons for using wood-ageing; however it must be stressed that by no means all wines benefit from prolonged contact with oak. A suitable wine must have reasonable balance in all its major components, be clear and have microbiological stability. Without these pre-requisites a wine may become even more unbalanced and spoilage and the build-up of solids in the wine and cask will be difficult to overcome and remove. Additionally, of course, only wines which are improved by the oak flavour and by ageing are placed in casks; other wines, no matter how well balanced and stable, are best aged in the bottle.

It is difficult to prescribe which wines are

suitable, but as a general rule tannic red wines and full-bodied whites may benefit. White wines are mostly aged for from three to twelve months and red wines for up to three years.

The types of wood used for making the cask also determine its usefulness. Oak is generally the most satisfactory and European or the cheaper North American species are mostly chosen. Each gives the wine a different character and the winemaker may wish to experiment with oak from different sources. In new casks a shorter period of time will be needed for wine to pick up its flavour.

The biochemical complexity of the ageing process in casks tests the skill of the winemaker. His role is to determine when the character of the wine is sufficiently complemented by the oak flavour, and yet not overpowered. His only tools are the tasting glass and his palate. The art cannot be learnt from books and a methodical approach and a will to experiment will be of great assistance. The tasting of wines in casks must be regular, and detailed notes should be made so that tastings can be adequately compared. Where possible, all casks should be tasted, since spoilage can develop in some and not in others.

The care of casks has been described in a previous chapter. Casks with wine are best stored in a cool, dry and stable environment and sealed with a silicone bung. The two basic positions of casks in storage racks are shown in Figure 15.1. In position A, the casks, after filling, are rotated so that the sealing-bung is covered by wine and loss through evaporation is reduced. In position B, the casks stand upright which allows for easier sampling, but the wine must be kept topped-up with wine of the same type kept separately in an inert container. Under warm conditions this may be up to once a week, and as considerable labour is involved when a large amount of wine is aged for periods of months or longer, storage position A is preferable. When using this storage position, samples are obtained by loosening one of the staves in the cask head. A special tool is inserted between the staves on the lower side of the head and briefly twisted to allow a jet of wine to be released.

The following spoilage problems may be encountered with wines stored in casks:

1. Oxidation. Initially a slight browning in colour is noticed, while later a bitterness is apparent in the flavour. The first stage can be arrested by adjusting the sulphur dioxide level and carefully topping-up. More advanced oxidation can only be corrected by blending with a larger amount of the same wine, or by using a colour removing substance such as charcoal or PVPP.

2. Bacterial spoilage. This is usually caused by bacteria, such as *Mycoderma vini*, growing on the surface of wine and producing an off-white film, or bacteria in the wine itself, such as *Acetobacter, Lactobacillus Streptococcus, Bacterium mannitopeum*. Initially a slight acetic acid (vinegar) odour is detected, while later more complex and unpleasant odours are found. These may be described as "mousy" or "musty", depending on the type of bacterial spoilage. Poorly-maintained wines and casks are often the cause, and the effect may derive from increased pH, lack of free sulphur dioxide and the presence of oxygen.

It is better to treat the infected wine separately and not to blend with sound wine to cover up faults. A slight off-odour may be overcome by aeration—achieved by pumping wine into containers without inert gas in such a way that the wine will absorb a large amount of air—followed by sulphiting and filtration. In more advanced stages this spoilage can sometimes be corrected by filtration through pads pre-

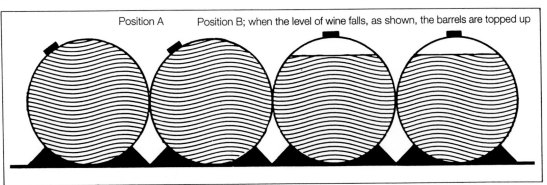

Position A   Position B; when the level of wine falls, as shown, the barrels are topped up

Figure 15.1 Maturation of wine in casks

packed with activated charcoal, followed by sulphiting and the destruction of bacteria by pasteurisation.

3. Yeast spoilage. The wine has a cloudy appearance and there may be a release of carbon dioxide. Wines with residual sugar of 5 g per litre or more can be subject to this spoilage. The wine should be removed from the barrel, fermented dry—less than 2 g per litre of sugar — restabilised, cleared and returned to cleaned barrels, which have been previously steamed.

4. Woodiness. This condition is caused through excessive ageing in casks. The wine should be removed from the cask and blended with less matured wine. Regular tasting avoids this problem.

Recent work by Peynaud in Bordeaux confirms previous observations that red wines, which absorb high levels of tannins from grape skins during fermentation, are less likely to pick up excessive oak flavours, especially from new barrels. He found with well-ripened Cabernet grapes that delayed pressing (20 or more days on skins) increased tannin levels and therefore reduced undesirable wood flavours. Thus 'over-oaking' or woodiness is usually a problem with wines low in grape-skin tannin.

5. Mercaptans and hydrogen sulphide. Off-odours caused by these sulphur compounds are typified by the smell of organic decay. It is caused by the action of yeast on certain must constituents. Some yeasts are more likely to cause this problem than others.

The following materials or conditions of the must are likely to predispose them to sulphur off-odours: high residues from sulphur sprays, high pH and a low nutrient level. Thus to reduce the problem a winemaker should use a known low-sulphur-producing yeast, be cautious in the use of sulphur sprays and in the use of sulphur dioxide in winemaking, correct the pH if required and adopt the addition of ammonium phosphate. If these odours are already present, the addition of copper sulphate will help — for details see Chapter 11.

The maturation of wines in bottles

Once the wine has gained the optimum benefit from cask maturation, or has been kept in an inert container long enough for the winemaker to be sure that it is sound and stable, it is bottled. Bottled wine may then be directly supplied to the merchant, or it may be bottle-aged for a further period by the winemaker who can then ensure that it reaches and is enjoyed by the consumer at a satisfactory level of maturity. The cost of cellar-ageing will hopefully be off-set by a higher price.

If wine is sold before reaching its full maturation potential, it gives the consumer a cheaper product with the option of cellaring the wine at home, thus gaining the pleasure of following each wine's development during bottle-ageing. This also reduces the winemaker's need for storage space and results in cost savings.

Bottled wines should be stored in a stable, dry, dark and cool atmosphere, free from excess humidity or vibration. Corked bottles are kept on their sides to keep the cork moist and prevent spoilage. Winemakers and merchants who store large quantities of wines taste randomly-selected, sample bottles periodically to assess the wine's progress and to determine the best time for selling. Records of such tastings are kept for future reference, and at least one tasting using a four-to-six member tasting panel may prove useful in comparing vintages and brands.

It is recommended that a person wishing to establish a home cellar should adopt a similar approach. It is preferable to buy a significant quantity of the same wine so that it is possible to follow closely the development of the wine at various stages of maturity. Keeping records of a

Figure 15.2 Commercial wine cellar at Bernkastel in the Moselle.

wine's price, source of origin and progress at home tastings should also prove useful and enjoyable.

The length of time required for bottle ageing depends on a number of factors; among them are winemaking techniques, the wine style required and, of course, the differing taste preferences of consumers. The following points should be noted and can be used as a general guide for selecting and storing wines.

1. Wines selected for ageing should be sound, clear and well balanced in their major components (alcohol, flavour, acidity, bitterness etc.). However, the actual amount of each component will change over a period of time and may make a slightly unbalanced wine balanced.

2. It must be understood that with ageing wines lose one set of characteristics and gain others. Thus freshness is replaced by complexity; acidity and slight bitterness change with time to mellowness and softness. The degree of these changes will be determined by the length of maturation. People of course have different taste preferences and this will also modify the time of ageing.

3. Wines stored under warm temperatures will mature a little faster than those in cooler cellars; 10-15°C (50-59°F) is considered the best range.

4. Good white and rose wines may be sufficiently matured in three to five years. However, some of the finer German and French wines, or their equivalents in countries outside Europe—for example German ausleses, French white-burgundies or Californian, Australian and New Zealand Chardonnay wines — may improve after a much longer period depending on vintage, wine style and winemaker. Sparkling wines, including French champagnes, are bottled ready for drinking and prolonged storage of these can be risky.

5. Red wines present even greater variety and scope for ageing; from the lighter styles — where three to five years is sufficient—to fine Cabernet, Pinot noir, Shiraz and others in the claret or burgundy style, where more than a decade can be required to bring a bottle to its best. Such types are the better wines from Bordeaux, Burgundy, the Rhône, Australia, California and New Zealand.

In general terms ageing of wines is a consequence of a series of physical, chemical and microbiological changes in which oxygen acts as a catalyst.

Quality control and sensory evaluation

Behind all the operations so far described is the requirement for quality control: to produce a wine as free from faults as is possible, while still retaining its basic character. This is the *science* of winemaking. The art of winemaking is more complex and difficult to describe, since it involves the ability to use sensory abilities to complement technical expertise. It also relies upon the taster's knowledge of what are the characteristics expected in the region, the variety and the style. This art must be learnt and what follows is a description of the background on which it can be built.

To evaluate wines, three senses are used—sight, smell and taste.

Sight

This is the least dependable of the senses as it varies considerably between tasters. Fortunately for the winemaker, colour and clarity evaluations can be simplified, and much variation may be eliminated by using colour standards selected to suit the public expectations for each wine style (claret, rose, etc.). Wines should also be clear, even brilliant, in appearance and free from suspended solids. A standard tasting glass of clear and thin glass should be used (see Figure 15.3).

Tasting rooms need to be equipped with a standard source of light as indirect light from outside is too variable. Some tasters prefer a candle flame as the light source since this obviously gives a similar amount of light anywhere in the world. It is important to have a white background such as large sheets of white blotting paper, or even painted tasting-tables.

Wines showing dullness or cloudiness should have marks deducted (see Tables 15.1, 15.2 and 15.3 later in this chapter), as should those with an inappropriate colour for the style being evaluated. However, it should be remembered that changes in colour will occur in older wines. White wines will become golden or even slightly amber with age, red wines will gain a brown hue; the colour of both will appear less brilliant and deeper, and small crystalline deposits can form. These changes are a natural part of the wine's maturation and should be allowed for.

Smell

Smell is the most important and complex of the senses used during wine evaluation, a fact often not recognised by less-experienced tasters. Smells are perceived through the olfactory mechanism situated in the upper-most section of the nasal passage. The air stream which carries the volatile components from wine enters the nostrils and is directed upwards into the 'olfactory bulb' by 'buffer plates'. The small cavity, known as the 'olfactory bulb', is the site of our sense of smell. This area is also connected by a small channel to the top of the back palate in the mouth, which means that the sense of smell is stimulated by both the air taken through the nose and by vapours coming from the wine held in the mouth. The latter sensation, however, is often regarded as a part of taste.

Tasting glasses should be tulip-shaped in order to concentrate the vapours, and if the wine is swirled in the glass prior to lifting to the nose, the bouquet is further concentrated. Glasses should be filled to one-third. Some types of tasting glasses are shown in Figure 15.3.

A few repeated sniffs of moderate intensity are better than excessively gentle or vigorous

Figure 15.3 Styles of tasting glasses.
All of these are satisfactory except the centre one This is too shallow and will allow the aroma to disperse before sniffing.

inhalations. Likewise, prolonged sniffing of a wine leads to confusion and reduced accuracy. For this reason a limit of 30-40 seconds per wine is more than sufficient to evaluate the bouquet. If several wines are to be evaluated, a gap of one to two minutes between smelling them should be left to avoid confusion and fatigue, and also to allow for a full recovery of the olfactory senses. Should any of the wines contain a dominant aroma or odour, for example the varietal character of Muscats or Gewürztraminer, or sulphur and volatile acidity, an even longer period of time should elapse between the smelling of different wines.

Winemakers and tasters sometimes refer to two components when defining the smell of a wine: the aroma, which comes from the varietal and regional characterstics, and the bouquet, which arises from fermentation and maturation.

Taste

There are five major taste sensations which will be described separately. These are: acidity-sourness, sweetness, saltiness, bitterness and touch or feel.

1. Acidity-sourness. Sourness is perceived by taste buds on the sides of the tongue. Common wine-acids have a different intensity or sharpness in flavour. Malic and citric acids are relatively sharp, whereas tartaric and lactic

acids appear softer in taste. It will be recalled that in cool-climate wine regions, where the malic acid content in the less-ripe grapes is high, the wine shows increased sharpness and even austerity. In hot-climate regions the reverse is true and wine often lacks freshness and fruitiness due to lack of acidity. During ageing, acids react with alkaline and alcoholic components to form salts (e.g. malates and tartrates) and esters. Thus the aged wine tastes less acidic. A small amount of carbon dioxide remaining in the wine for a short time after bottling can increase the apparent acidity of young wines.

2. Sweetness. This sensation is mainly due to residual sugars, although a highly-flavoured or alcoholic wine will appear sweeter than the amount of sugar present would suggest. Lack of acidity can also increase the apparent sweetness in wine, whereas lack of body (alcohols, extract) will have the opposite effect. High apparent sweetness, such as are found in German ausleses or French Sauternes wines will decline with bottle age due to loss of primary aroma.

Sugars are most accurately perceived by taste buds on the tip of the tongue, though in sweet wines they are sometimes apparent throughout the palate and in the aftertaste. As with acidity, sweetness is sometimes a required and basic component of the wine style; though generally it should not be directly obvious, and it should complement rather than overpower all the other components of wine flavour.

3. Bitterness. Bitterness in wine is caused by tannins and other phenolic substances. In wine, most originate from grape skins, stalks, seeds and from the oak in cask-fermentation or maturation. Bitterness is perceived by taste at the back of the tongue and, to some extent, on the sides of the mouth: it can be particularly conspicuous in young, red wines. With ageing the tannins are chemically changed to compounds which are no longer so bitter or astringent, and the mellow taste of matured, red wines demonstrates this clearly.

Astringency is the mouth-puckering feel of young red wines, obvious on the back palate as opposed to sourness on the front palate. As such, tasters refer to astringency as a component of the wine's texture, e.g. 'feel', rather than flavour structure — to which bitterness belongs .

4. Saltiness. Saltiness is rarely found in wines, except in the 'flor' type fino sherries of Spain or,

to a lesser extent, the Jura whites of France. It is detected by taste buds of the middle palate on top of the tongue.

In some white wines a high metal content (iron or copper), if present in excessive amounts, can create an impression of saltiness. As this cannot be said to form a natural part of wine flavour it should be regarded as an undesirable characteristic.

5. Touch or feel. Although not always recognised as one of the basic taste sensations, touch and feel form an important part in wine evaluation. The feeling of fullness in the mouth, also called weight or body, is mainly due to alcohols, sugar, astringency, extract and other components present in the wine, sometimes in minute quantities. Lack of these can make a wine feel thin or even watery.

A level of pain can be associated with the taste of wine: this is especially obvious when tasting wines with excessive acidity, bitterness or sulphur.

Other considerations

There are a number of other factors which can have an effect on a taster's ability to assess wines correctly and consistently. Noise, visual distractions, comments by fellow tasters or their facial expressions can all contribute to misleading results. A person's expectations, a lack of ability to concentrate, or a lack of confidence in one's own judgements are common among inexperienced tasters. Specially designed procedures* for selecting taste-panels may be used for serious evaluations. A good tasting-room design, adequate training of potential tasters and careful analysis of results—using statistical procedures if needed — provide a sound base for sensory wine-evaluation.

It is also important that tasters do not feel tired and tasting sessions should be timed to avoid fatigue — the late morning hours are most suitable. Where comparative evaluations are required, a panel of experienced tasters will prove more reliable than a single taster, no matter how well qualified. A panel of five to seven tasters is considered satisfactory.

Despite the wide-spread belief that expert wine tasters are born, not made, most people with sufficient training and experience can qualify for tasting panels.

---

* Amerine, M.A. and Roessler, E.B. 1976. *Wines—their Sensory Evaluation.* Freeman and Co., San Francisco.

The purpose of wine evaluation

Wine may be evaluated for a number of reasons.

1. Routine evaluation by the winemaker during production.

A winemaker, with or without his associates, will taste wines at different times during fermentation, storage and maturation to assure himself that it is sound, free from off-odours or off-flavours, and developing the characteristics desired for a given style.

Where wines are being blended or adjusted, the winemaker may present his final choice to a panel of tasters for evaluation. Wines or their blends are then tasted blind—without any means of identification, often using the triangular method. In this the wine is presented alongside two standard wines, or previous blends, and the taster's preference noted, this is repeated several times, using different combinations. A consistent preference by the tasters indicates the desirability or not of the adjusted or blended wine.

2. Routine evaluations by merchants when buying or ageing wines.

For marketing, wines can be placed in two basic groups: bulk wines and the more prestigious varietal or regional wines.

Bulk wines. The more successful wines of this type will be consistent in quality, have a high public acceptance of their style, and a suitable price range. Public tastings of wines with a similar style can be useful in establishing market preferences.

Varietal or regional wines. These wines are produced for a specialised section of the market and distinction in character coupled with individuality in style are important. Variation in price and lack of consistency between vintages are secondary considerations. Public tastings are also less important than with bulk wines and the merchant will be guided by his own assessment which could result from several tastings of each wine during maturation. For such tastings, wines are served 'blind' — without means of identification—either singly or in groups, according to variety, vintage, district etc.

3 . Wine exhibitions.

Tastings of wines at exhibitions are popular in all wine-producing countries and have a two-fold value. Firstly, they attract publicity to the better wines and serve as a public relations exercise. Secondly, the results and comments of judges are useful to winemakers as an objective evaluation of their wines alongside those of other competitors.

Wines at exhibitions are grouped according to age, style or grape variety, and are tasted by panels of up to ten judges. The tasting forms used vary in minor details, though the majority specify the maximum points which can be gained for colour, bouquet and flavour, and give the minimum number of points required for various awards. Although wines are judged individually, the number of wines (often hundreds per class) may necessitate some screening first: the elimination of lesser wines and those with faults assists the evaluation of the better ones.

Unfortunately the evaluation of a large number of wines often results in judges choos-

Table 15.1 Form A. Tasting form for separating varieties into groups according to quality, and eliminating those not worthy of further evaluation.

Taster			Date								
Wine											
			Place your score for each wine in the appropriate column								
Scoring	Points		Wine Number								
		1	2	3	4	5	6	7	8	9	10
Unacceptable	1										
Average quality with faults	2 3										
Average quality	4 5 6										
Above average with single outstanding characteristic	7 8										
Superior	9 10										

Wines scoring 7 points or above are further evaluated in a more detailed tasting session using Form B.

ing those with obvious characteristics rather than those with more subtle ones. This is especially the case where wines of different ages or varietal types are grouped together. In excessively large classes of wines, or in groups which are more demanding such as young red or fortified wines, tasting fatigue can be a serious problem.

4. Specific evaluations for research purposes.

These are used to assess new varieties for established districts, or to select varieties for new grape-growing areas. Wines which have been produced by using new techniques of vinification or wine handling may also be compared with those already in use. When assessing varieties, the first consideration is the presence and balance, or absence, of distinctive aroma and flavour characteristics. Winemaking techniques should be standardised and not complicated, and the assessment of the results should be made with the awareness that a commercial winemaker would have the ability to ameliorate any minor imbalances, or a create more complex style.

5. Tastings to establish the wine's origin, identity, and the grape variety used.

This can be a somewhat dilettante operation, employed for its entertainment value or, at a more serious level, for educational purposes.

It demands for the taster a wide understanding of grape characteristics and their geographically and regionally associated wine styles, coupled with a sound knowledge of winemaking techniques.

This, when combined with extensive tasting experience and memory—aided by detailed tasting notes — can lead to interesting and rewarding sessions.

Wines can be tasted 'blind' and the taster can be asked to identify them with clues, such as the variety, style, or district of origin, e.g. the 'options games' popular in Australia and New Zealand.

Tasting forms and terminology

If a tasting panel is used it is important to ensure that all members are using the same terminology and have similar standards of evaluation. The use of forms such as those shown in on the following pages help to provide uniformity of assessment. When evaluating a wine it is correct to begin with the top marks available for each category and then to subtract marks if it does not meet the standard expected. The comparative value of a wine is then given by the total points scored averaged over all judges.

Tasting sessions should be timed in such a manner that they occur over the full period of a wine's life. It is then possible to assess the wine at its peak of full maturity and to note the stages of its eventual decline.

Table 15.2 Form B: Detailed tasting form for general evaluation of wine, or for evaluating individual varietal characteristics

Name: ....

Date: ....

Notes for tasters:

(a) Evaluate each parameter in chronological order
(b) Subtract points from the maximum where you feel there is a definite reason not to award full points
(c) The adjectives in the motivation column list the range of each parameter to be considered.
(d) Scoring intervals of 0.5 points may be used

Parameter	Maximum Points	Motivation		Wine Number					
				1	2	3	4	5	6
A. Condition (1)	1	Brilliant	1						
		Clear	0.5						
		Dull/Cloudy	0						
B Colour (2)	2	Excellent, good depth	2						
		Satisfactory, fair depth	1						
		Undesirable and/or lacking	0						
C Aroma and Bouquet (6).									
Intensity and persistence of aroma and bouquet	4	High	4						
		Moderate	2						
		Low	0						
Softness, richness and complexity	2	High	2						
		Moderate	1						
		Low/lacking	0						
Foreign spoilage odours	2 (Subtract)	High	-2						
		Moderate	-1						
		Not present	0						
D Taste-Flavour (11)									
(i) Sourness (acidity)	2	Pleasant, balanced	2						
		Slightly unbalanced	1						
		Lacking or excessive	0						
(ii) Fullness and body	1	Correct for style and type	1						
		Too alcoholic or thin and watery	0						
(iii) Intensity and persistence of desirable flavour	3	High	3						
		Moderate	1.5						
		Low	0						
(iv) Bitterness	1	Soft and rounded	1						
		Too bitter or too flat	0						
(v) Aftertaste of desirable qualities	2	Very good, lasting	2						
		Good	1						
		Poor	0						
(vi) Overall impression/harmony	2	Excellent	2						
		Good	1						
		Poor	0						

Table 15.3 Form C: Tasting form for evaluation of wines with common characteristics (e.g. yield, time of maturity, alcohol level). Groups normally have between three and six wines.

Group and wine number		Points for each wine*				Order of preference			Comments
		Colour	Smell	Taste	Total	Colour	Smell	Taste	Overall impression
Group A	Wine No.								
	1								
	2								
	3								
	4								
	5								
	5								
	6								
Group B	Wine No.								
	1								
	2								
	3								
	4								
	5								
	6								
Group C	Wine No.								
	1								
	2								
	3								
	4								
	5								
	6								

*Groups are sampled separately. Points scored are from 0 (poor) to 3 (good) and the total in column 4 gives a comparative assessment of the wines. Columns 5 6 & 7 give an order of preference for colour smell and taste

CHAPTER 16
THE FUTURE

For well over six thousand years grapes have been grown and wines have been made throughout the Middle East and Europe. Initially, attempts to produce wine must have been very haphazard and more often than not the resultant brew would have been better vinegar than wine. Slow improvement in techniques, however, would gradually have increased the chances of success, although it is doubtful whether the quality produced would have matched our present standards in all but a few instances.

In any district, different types of grape vines would have become recognised, and the better ones, over long periods of time, would have become dominant. Slowly a reputation might have become established, especially where the variety, the soil and the climate were ideally matched, and where the local winemakers had developed an appropriate level of skill. The Rhine and Mosel, Bordeaux and Burgundy, have reputations gained slowly and laboriously over centuries. They are not likely to be suddenly lost, but new reputations will be made in other areas much more quickly. The already-established areas will always produce distinctive wines, but their preponderance as the premium world wines will be challenged.

Technological improvements in viticulture and oenology are now widely publicised — often within six months of the research which brought them about—and can be responsible, within a year or less, for changes in a wine cellar some 12 000 miles away. New chemicals for modifying vine growth may be distributed simultaneously throughout the world by one chemical company keen to promote its particular product. A new grape variety may have taken some twenty years to develop and assess at a research institute, but, within a few years, research stations through out the world may be propagating this material for evaluation in local conditions, and for distribution to selected growers. The idea of a 'global village' is entirely pertinent to this industry, and indeed, it might be easier to transfer knowledge across the world in the twentieth century than it was between two villages in the twelfth.

These global changes will, by the end of the century, have entirely altered the reputation of wines from certain countries. For example, in New Zealand, wines up to twenty years ago were very mediocre and it was difficult to persuade connoisseurs to drink the local product. To imagine that New Zealand wines would be on the shelves in overseas countries was almost unthinkable. Now, with the discovery that many classical varieties can, with the aid of modern therapeutant sprays, be grown in the wetter parts of New Zealand's North Island; that some South Island areas— which are drier and cooler—can also produce excellent wines; quality has improved beyond recognition and export is a reality.

In the eastern and north-western states of the U.S.A. and Canada similar improvements are occurring, and new cool-climate districts in Australia and Tasmania will begin to make their mark. This does not mean that the traditional wine-growing areas of Europe will be displaced — one cannot imagine anything finer than a German auslese or a top French claret—but other styles, some similar, some different, will appear and new areas will begin to be judged on the style they can consistently produce. Already non-European styles are in evidence, for example the Cabernet

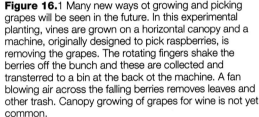

Figure 16.1 Many new ways ot growing and picking grapes will be seen in the future. In this experimental planting, vines are grown on a horizontal canopy and a machine, originally designed to pick raspberries, is removing the grapes. The rotating fingers shake the berries off the bunch and these are collected and transterred to a bin at the back ot the machine. A fan blowing air across the falling berries removes leaves and other trash. Canopy growing of grapes for wine is not yet common.

Sauvignon and Chardonnay wines of California, the Pinot noirs of Oregon, the Shiraz or Hermitage of Australia, and the Sauvignon blancs of New Zealand. Many of these are fine wines, consistently different and distinct from European wines and from each other. But the interesting fact is that they have been developed in decades and not centuries. We are in a fascinating era and those of us who have the good fortune to live in cool-climate, grape-growing areas will have the opportunity to observe these exciting developments within the course of a lifetime.

Nevertheless, rapid progress and success is not necessarily assured and grape-growers, winemakers, governments, and other institutions will need to guide this development with the utmost care. Success, we believe, will be largely dependent on the following:

1. The ability of the regional grape-grower and winemaker to match varieties plus techniques of growing and vinification to local climates and soils. Quality wines with distinctive style will result from the balance achieved between these factors. The speed with which this balance can be attained will be hastened —and the cost of development will be reduced —by organisation and co-operation between growers, and by the development of teaching and research facilities in each district. The Davis campus of the University of California, and the numerous research and teaching establishments in the various German wine districts, have had inestimable effects on the Californian and German wine industries. The sooner such facilities can be established, the greater will be their cost effectiveness.

2. The presence of sensible wine regulations aimed at the quality control of local wines and at their promotion. Such regulations might be three-tiered. The first stage, for example, could be the geographical delineation of areas; the second, correct labelling which truly reflects varietal composition, the use of natural vinification, and any special character —such as late harvest, berry selection and the presence of *Botrytis*. The third stage would be the establishment of a tasting panel of experts who would decide on the quality of wines for premium classes and on their suitability for export.

As local markets become saturated, produces will need to consider exporting their wines. It might be tactically advisable to aim exports at centres of population in South-East Asia, Japan, Canada and America rather than the saturated, traditional areas of Europe. Wine drinkers in Europe will be interested in tasting new wines from overseas, but these will be judged on the standards of Europe and they will probably never become a significant proportion of the total wine consumed. Asia and North America, are expanding markets with less tradition and no existing prejudices. A small increase in wine consumption would represent a large total volume, and exports to these areas could probably take all the surplus production of new wine districts.

It is to the growers and winemakers of such areas, and to those who will guide their progress and development, that we dedicate this book.

APPENDICES

Introduction

The appendices contain the analytical methods used by winemakers from the time of harvest to bottling. Each method includes, first, the reagents and equipment required, and second, a step-by-step procedure describing the method of analysis. The methods described are selected as being those most suitable for relatively untrained personnel and those appropriate for routine analysis in laboratories with limited rather than expensive and complex equipment.

Whilst the methods chosen are simple, any operator with no previous experience in a laboratory would be wise to consult someone who has such experience, especially as regards safety precautions. Some dangerous chemicals will need to be handled, and while safety information is included in the text this must, of necessity, be very basic and not totally comprehensive.

General comments

No two analyses of the same sample will yield exactly the same result. It is recommended, therefore, that all analyses be made in duplicate and that the average of the two results be taken and used. With time the likely variation will be determined and in those analyses where variations are small they might, subsequently, be made only once.

Sometimes reagents may be bought in packs or ampoules, with instructions on how to dilute them to arrive at specific concentrations. These are valuable, accurate and save time; but since their availability varies from country to country, preparations of reagents for all methods are described so the winemaker will have the choice of either method.

Chemicals should normally be of the quality described as 'Analytical Reagent' (AR) in contrast to the less pure 'Laboratory Reagent' (LR). Chemicals are referred to by name rather than by formula. Thus, if sodium hydroxide is named it signifies AR sodium hydroxide, i.e. AR NaOH. Distilled water should be used for all dilutions and also wherever water is specified in the appendices.

All reagents should be stored in glass containers—bottles or jars—properly sealed and stored in locked cabinets under cool and dry conditions. When reagents are made up to a recommended concentration they must be dated, since many will not keep longer than a period which will be specified in the text.

When diluting chemicals use the more accurate volumetric flasks rather than measuring cylinders. Glass should be heat-resistant Pyrex which can be placed over a naked flame and which will resist heat given off when mixing chemicals. Adding strong acids or alkalis to water, rather than the other way around, will reduce the heat which can be formed and, if this is likely to be great—as when sulphuric acid is added to water, the container should be placed in a cold-water bath.

Each laboratory should be equipped with protective clothing—laboratory coats, rubber gloves, goggles—and a first-aid cabinet. Basic methods for treatment of burns, eye damage and poisoning should be understood.

Appendix 1
Determination of sugar levels

Sugar levels above 2.2 °Brix (1 °Baumé)

Use of hydrometer

Transfer 100-250 ml of the sample into a 250 ml measuring cylinder, float a hydrometer in the sample and read the level of sugar, taking the figure just below the liquid level. It is important to avoid air bubbles in the sample, for they would give erroneous readings. Check that the temperature of the sample corresponds to the temperature at which the hydrometer has been calibrated. If it doesn't, adjust the reading using Table A.1.1.

Thus, if the temperature of the sample is 16°C and the °Brix reading 8%, the corrected figure is 8 — 0.19 = 7.81; if the temperature is 26°C and the Brix reading 21%, then the corrected level is 21 + 0.40 = 21.4.

Refractometer

The hand-held refractometer is useful for field sampling prior to harvest. A number of berries from different parts of the vine and bunch are selected and squeezed; a large drop of juice is then placed on the refractometer

---

* Note: °Brix more accurately measures soluble solids which, as Table A.1.4 shows, is slightly more than sugar.

glass. After closing the lid the sugar level*— usually in °Brix—is read directly on the scale. Temper-ature can also affect the refractometer reading and corrections should be made as in Table A.1.2.

Specific gravity

The specific gravity (SG) of a liquid is the weight of a known volume divided by the weight of an equal volume of water at a specified tempera-ture. The more dissolved sugar there is in a juice, the greater is the specific gravity. This can then be converted to sugar equivalents—see Table A.1.4. Hydrometers are sometimes cali-brated to read in specific gravity but these levels may be measured more accurately by a pycnometer. Weigh a pycnometer empty, with juice, and with distilled water to four decimal places,
then: Specific gravity =

$$\frac{\text{wt. with sample—wt. empty pycnometer}}{\text{wt. with water—wt. empty pycnometer}}$$

Temperature corrections if required are as shown in Table A.1.3

Generally, grape growers would use a refrac-tometer as a standard piece of equipment in the vineyard, winemakers would more commonly

---

Table A.1.1   Correction for Brix hydrometers calibrated at 20°C

Temperature of Liquid (°C)	° Brix reading in sample					
	0	5	10	15	20	25
10	-0.32	-0.38	-0.43	-0.48	-0.52	-0.57
11	-0.31	-0.35	-0.40	-0.44	-0.48	-0.51
12	-0.29	-0.32	-0.36	-0.40	-0.43	-0.46
13	-0.26	-0.29	-0.32	-0.35	-0.38	-0.41
14	-0.24	-0.26	-0.29	-0.31	-0.34	-0.36
15	-0.20	-0.22	-0.24	-0.26	-0.28	-0.30
16	-0.17	-0.18	-0.20	-0.22	-0.23	-0.25
17	-0.13	-0.14	-0.15	-0.16	-0.18	-0.19
18	-0.09	-0.10	-0.11	-0.11	-0.12	-0.13
19	-0.05	-0.05	-0.05	-0.06	-0.06	-0.06
20						
21	0.04	0.05	0.06	0.06	0.06	0.07
22	0.10	0.10	0.11	0.12	0.12	0.13
23	0.16	0.16	0.17	0.17	0.19	0.20
24	0.21	0.22	0.23	0.24	0.26	0.27
25	0.27	0.28	0.30	0.31	0.32	0.34
26	0.33	0.34	0.36	0.37	0.40	0.40

Table A.1.2   Correction for Brix refractometers calibrated at 20°C

Temperature of Liquid (°C)	° Brix reading in sample					
	0	5	10	15	20	25
10	-0.5	-0.54	-0.58	-0.61	-0.64	-0.66
11	-0.46	-0.49	-0.53	-0.55	-0.58	-0.60
12	-0.42	-0.45	-0.48	-0.50	-0.52	-0.54
13	-0.37	-0.40	-0.42	-0.44	-0.46	-0.48
14	-0.33	-0.35	-0.37	-0.39	-0.40	-0.41
15	-0.27	0.29	-0.31	-0.33	-0.34	-0.34
16	-0.22	-0.24	-0.25	-0.26	-0.27	-0.28
17	-0.17	-0.18	-0.19	-0.20	-0.21	-0.21
18	-0.12	-0.13	-0.13	-0.14	-0.14	-0.14
19	-0.06	-0.06	-0.06	-0.07	-0.07	-0.07
20						
21	0.06	0.07	0.07	0.07	0.07	0.08
22	0.13	0.13	0.14	0.14	0.15	0.15
23	0.19	0.20	0.21	0.22	0.22	0.23
24	0.26	0.27	0.28	0.29	0.30	0.30
25	0.33	0.35	0.36	0.37	0.38	0.38
26	0.40	0.42	0.43	0.44	0.45	0.46

use a hydrometer. Table A.1.4. shows the relationship between specific gravity, Baume, °Brix, sugar and potential alcohol.

Sugar levels below 2.2 °Brix (1°Baumé)

Lane and Eynon method

- **Reagents**

  Fehlings solution A Dissolve 34.6 g of cupric (copper) sulphate pentahydrate in 200 ml of water and make up to 500 ml.

  Fehlings solution B Dissolve 173 g of sodium potassium tartrate and 50 g of sodium hydroxide pellets in 200 ml of water and make up to 500 ml.

  Mixed Fehlings solution Mix well 50 ml of solution B and an equal volume of solution A.

  Methylene blue indicator Make up 1 per cent solution by dissolving 1 g of methylene blue in 100 ml of water.

  d-Glucose (dextrose) Dissolve 1.25 g of d-glucose in 50 ml of water, add 60 ml of pure ethyl alcohol and make up to 250 ml with water. This will keep for only two weeks.

  Note: Fehlings solutions A and B are stable for at least two months; the mixed solution is stable for only 24 hours.

- **Method**

  1. Standardising the mixed solution — Pipette 10 ml of the mixed solution into a 250ml Erlenmeyer flask containing 40 ml of water and a few glass beads. Boil for 15 seconds and add a few drops of methylene blue indicator. Titrate with d-glucose solution from a burette while still boiling until a definite red-colour persists. The entire titration should be carried out within 60 seconds. The number of ml of d-glucose used in the titration is the standard of the mixed Fehlings solution. Record as 'Volume A'.

  2. Boil 20 ml of the sample for 20 seconds, cool and pipette 10 ml into a 250 ml Erlenmeyer flask containing 40 ml of water and a few glass beads. Add 10 ml of the mixed solution and bring to the boil. Add a few drops of methylene blue indicator and boil for 15 seconds.

  3. Titrate the boiling sample with d-glucose from a burette until all the blue colour is removed and a definite, red end-point persists. The number of ml used in this titration is recorded as 'Volume B'.

  4. Calculate the amount of sugar in the sample as follows: (Vol. A—Vol. B) x 0.5 = sugar in grams per litre.

- **Notes** Wines with more than 10 g per litre (1 per cent) of residual sugar should be diluted to one tenth (1ml sample in 9 ml water). The result from step 4 above must then be multiplied by 10. An indication of a higher level of residual sugar is given by the inability of the mixed Fehlings solution and methylene in-

Table A.1.3 Correction for specific gravity hydrometers calibrated at 15.6°C

Temperature of Liquid (°C)	Specific gravity reading in sample				
	0	0.2	0.4	0.6	0.8
10.0	-0.0015	-0.0014	-0.0014	-0.0013	-0.0013
11.0	-0.0012	-0.0012	-0.0011	-0.0011	-0.0010
12.0	-0.0010	-0.0009	-0.0009	-0.0008	-0.0008
13.0	-0.0007	-0.0007	-0.0006	-0.0006	-0.0005
14.0	-0.0004	-0.0004	-0.0003	-0.0003	-0.0002
15.0	-0.0002	-0.0001	-0.0001	0.0000	0.0000
16.0	0.0001	0.0001	0.0002	0.0002	0.0003
17.0	0.0003	0.0004	0.0004	0.0005	0.0005
18.0	0.0006	0.0006	0.0007	0.0007	0.0008
19.0	0.0008	0.0009	0.0009	0.0010	0.0010
20.0	0.0011	0.0012	0.0012	0.0013	0.0013
21.0	0.0014	0.0014	0.0015	0.0015	0.0016
22.0	0.0016	0.0017	0.0017	0.0018	0.0018
23.0	0.0019	0.0019	0.0020	0.0020	0.0021
24.0	0.0021	0.0022	0.0022	0.0023	0.0023
25.0	0.0024	0.0024	0.0025	0.0025	0.0026

dicator to turn the sample blue prior to the titration with d-glucose.

Red wine samples will need first to be decolourised by adding 0.5 g activated charcoal, mixing well and filtering through Whatman No. 1 filter paper.

Dextro-check method

'Dextro-check' or 'Clinotest' kits are used by diabetics to test levels of sugar in their urine. They are just as suitable and convenient for determining low levels of sugar in wine. One tablet is dissolved in a 0.5 ml wine sample using the test tubes provided. The level of sugar is read directly from a colour chart. This method gives an approximate result, however it is recommended for routine sugar checks in cellars, being simple and less time consuming than methods which would be necessary for more exact assessments.

Combi Test

This is a sensitive method for determhling sugar in wines. The method is too detailed to be described here but it has many advantages, one of which is that the same apparatus can be used to measure alcohol. Readers are referred to Van Dam, T.G.J. 1979. 'A Manual of Basic Laboratory Methods for Winemakers'; Oenological and Viticultural Bulletin No. 11. Ruakura Soil and Plant Station, Private Bag, Hamilton, New Zealand.

Appendix 2
Determination of total acidity

- **Reagents**

0.133N sodium hydroxide* Dissolve 5.32 g of sodium hydroxide pellets in 500 ml of water. Make up to 1 litre with water. Check strength (normality) with potassium hydrogen phthalate solution—see below.

Potassium hydrogen phthalate Accurately weigh 0.7 g** of potassium hydrogen phthalate and dissolve in 50 ml of hot water. Make this solution fresh for each check.

Phenolphthalein indicator Dissolve 1 g of phenolphthalein in 50 ml of pure ethyl alcohol and dilute to 100 ml with water.

Checking the normality of sodium hydroxide Add a few drops of phenolphthalein indicator to the 50 ml of potassium hydrogen phthalate solution and titrate with the sodium hydroxide

---

* 0.1N sodium hydroxide may be purchased in solution form. Its normality will still need to be checked as above and a *conversion factor* used on the final figure.

**It is not essential that 0.7g phthalate is weighed. The key factor is that the weight is accurate and this is used in the normality calculation.

Table A.1.4. The relationship between specific gravity, Baumé, °Brix (Balling), per cent sugar and potential alcohol

Specific Gravity at 15°C	Baumé	Brix or Balling	Sugar (per cent. w/v)	Potenfial Alcohol (per cent, v/v)
1.050	6.9	12.4	10.3	6.0
1.051	7.0	12.6	10.6	6.2
1.052	7.1	12.7	10.8	6.3
1.053	7.2	12.9	11.1	6.5
1.054	7.4	13.3	11.4	6.7
1.055	7.5	13.5	11.6	6.8
1.056	7.6	13.7	11.9	7.0
1.057	7.8	14.0	12.2	7.2
1.058	7.9	14.2	12.4	7.3
1.059	8.0	14.4	12.7	7.5
1.060	8.1	14.6	13.0	7.6
1.061	8.3	14.9	13.2	7.8
1.062	8.4	15.1	13.5	7.9
1.063	8.5	15.3	13.8	8.1
1.064	8.6	15.4	14.0	8.2
1.065	8.8	15.8	14.3	8.4
1.066	8.9	16.0	14.6	8.6
1.067	9.0	16.2	14.8	8.7
1.068	9.2	16.5	15.1	8.9
1.069	9.3	16.7	15.4	9.0
1.070	9.4	16.9	15.6	9.2
1.071	9.5	17.1	15.9	9.3
1.072	9.7	17.4	16.2	9.5
1.073	9.8	17.6	16.4	9.6
1.074	9.9	17.8	16.7	9.8
1.075	10.0	18.0	17.0	10.0
1.076	10.2	18.3	17.2	10.1
1.077	10.3	18.5	17.5	10.3
1.078	10.4	18.7	17.8	10.5
1.079	10.5	18.9	18.0	10.6
1.080	10.7	19.2	18.3	10.8
1.081	10.8	19.4	18.6	10.9
1.082	10.9	19.6	18.8	11.0
1.083	11.0	19.8	19.1	11.2
1.084	11.1	20.0	19.4	11.4
1.085	11.3	20.4	19.6	11.5
1.086	11.4	20.5	19.9	11.7
1.087	11.5	20.7	20.2	11.9
1.088	11.6	20.8	20.4	12.0
1.089	11.8	21.2	20.7	12.2
1.090	11.9	21.4	21.0	12.3
1.091	12.0	21.6	21.2	12.5
1.092	12.1	21.8	21.5	12.6
1.093	12.3	22.1	21.8	12.8
1.094	12.4	22.3	22.0	12.9
1.095	12.5	22.5	22.3	13.1
1.096	12.6	22.7	22.6	13.3
1.097	12.7	22.8	22.8	13.4
1.098	12.9	23.2	23.1	13.6
1.099	13.0	23.4	23.4	13.8
1.100	13.1	23.6	23.6	13.9
1.101	13.2	23.7	23.9	14.0
1.102	13.3	23.9	24.0	14.2
1.103	13.5	24.2	24.2	14.4

Note The German scale of Degree Öchsle (°Ö) is based on the difference ln weight by which 1 litre of must is heavier than 1litre of water. The first three figures, after the decimal point, of the specific gravity are equivalent to Öchsle—e.g. 1.075 S G = 75 °Ö; 1.101 S G = 101 °Ö.

from a burette until the pink end-point persists for 15-20 seconds.

Then normality =

$$\frac{\text{Weight of phthalate** } \times 1000}{\text{Number of mls sodium hydroxide} \times 204.22}$$

If the normality of the sodium hydroxide is outside the acceptable range of 0.131 —0.135 use a *conversion factor* to adjust the results of the titrations. Thus:

If normality = 0.137 multiply the result by
$\frac{0.137}{0.133} = 1.03$
If normality = 0.120 the conversion is
$\frac{0.120}{0.133} = 0.90$

• **Method**

1. Pipette 10 ml of must or wine sample into a 250 ml Erlenmeyer flask containing about 40 ml of water.
2. Add a few drops of the phenolphthalein indicator and titrate with sodium hydroxide from a burette until the pink end-point persists for 15-20 seconds.
3. If using 0.133N sodium hydroxide the number of mls of sodium hydroxide used is

equivalent to the number of grams of tartaric acid equivalents in the sample. If the normality is outside the acceptable range use the *conversion factor* as shown above.

• **Notes**

Red-wine samples should be de-colourised by shaking with approximately 0.5 g activated charcoal and filtering through a Whatman No. 1 filter paper.

Where available, a pH-meter can be used to indicate the titration end-point in place of phenolphthalein. If this method is used then titrate directly into a 100 ml beaker containing the diluted sample in which the tip of the pH-meter electrode is submerged. A pH of 8.1 represents the end-point. Using the pH-meter is a more accurate method for determination of acidity and red wines do not need to be de-colourised.

When analysing fermenting musts or young wines it is necessary that they first be de-gassed. This can be done by bringing the sample to the boil for 30 seconds. It will be ready for analysis when cooled to its original temperature. See Appendix 3 for instructions on the use of a pH-meter.

** See footnote on page 173

Appendix 3
Determination of pH

Litmus paper

As a routine during cellar work, litmus paper strips can be used to determine the approximate pH level. Strips can be bought ready to use with a colour chart in the packet. Different combinations of colours indicate the pH of the sample.

pH-meter

The most accurate method of measuring pH is by the use of a pH-meter, which will usually be the first item of expensive laboratory equipment a winemaker will buy. Instructions for its proper installation and correct use will be supplied on purchase.

• **Reagents**

Buffer solution, pH 3.5 Prepare saturated solution of potassium hydrogen tartrate. Do not store longer than seven days.

Buffer solution, pH 4.0 Dissolve 10.21 g of potassium hydrogen phthalate in 500 ml of hot water and dilute with water to 1 litre.

• **Method**

1. Calibrate the pH-meter using each of the buffer solutions until both give the correct reading on the scale. Wash the electrodes with water and check that the temperature of the wine sample is the same as that at which the pH-meter is set.
2. Pour 50 ml of sample into a small beaker and insert electrodes, but do not allow them to touch the sides or bottom. Switch the pH-meter to the 'on' position and record the pH level on the scale.

Appendix 4
Determination of sulphur dioxide

Two procedures — the 'Ripper' and the 'Aspiration' method—can be used to measure both free and bound sulphur dioxide. The Ripper Method should prove adequate for routine checks, although it tends to give a slightly higher reading than the correct one. The Aspiration Method, though more time consuming and complex, is recommended for the final check prior to bottling.

Free sulphur dioxide is the effective anti-oxidant and preservative and its level is of more significance to the winemaker.

The Ripper method

• Reagents

Starch indicator Buy 'ready to use' or boil 1g of starch in 100 ml of water for three minutes. Allow to cool before use.

25 per cent sulphuric acid Pour, slowly, 500 ml of concentrated sulphuric acid into a 5-litre Pyrex beaker containing 1.5 litres of water. Allow to cool before use.

10 per cent sodium hydroxide Dissolve 100g of sodium hydroxide pellets in 0.75 litres of water in a Pyrex glass beaker. Allow to cool and dilute to 1 litre with water.

Standard 0.1N iodine solution Dissolve 24g of potassium iodide in 150 ml of water, add 12.7g of iodine and dissolve. Make up to 1 litre with water. Store in brown bottles in a cool dark place; this is stable for at least four months.

0.02N iodine Dilute 400 ml of 0.1N iodine solution to 2 litres with water. This is stable for four weeks in a cool and dark position.

0.02N sodium thiosulphate Dissolve 1.241 g sodium thiosulphate pentahydrate in 50 ml of hot water. Allow to cool and dilute to 250 ml with water. This will keep stable for four weeks.

• Notes

Remember that the mixing of sulphuric acid and water is dangerous and must be done carefully — goggles are recommended when using this acid.

Concentrated solutions of sodium hydroxide (also dangerous) should be kept in a glass jar with a rubber, not glass, stopper.

Checking the normality of 0.02N iodine

Add 2-3 ml of starch indicator to 50 ml of sodium thiosulphate solution in a 250 ml flask. Titrate with 0.02N iodine from a burette until the purple colour persists for 15-20 seconds. Record the amount of iodine (the 'titre')

$$\text{Normality} = 0.02 \times \frac{50}{\text{mls of titre}}$$

• Method

1. Transfer by pipette a 50 ml wine sample into a 250 ml Erlenmeyer flask containing about 50 ml of water.
2. Add 10 ml, 10 per cent sodium hydroxide, and allow to stand, stoppered, for 15 minutes.
3. Add 10 ml, 25 per cent sulphuric acid, and 2-3 ml of starch indicator and titrate with 0.02N iodine from a burette until the purple colour persists for 15-20 seconds.
4. The number of mls of iodine used in titration, multipled by 12.8, is equivalent to the ppm (parts per million) of total sulphur dioxide in the sample. Example: titration 8.5 ml then: 8.5 x 12.8 = 108.8 ppm total sulphur dioxide.

If the normality of iodine, as checked above, is not 0.02, multiply the ppm total sulphur obtained by $\dfrac{\text{correct normality}}{0.02}$

Example: correct normality of iodine 0.025

Sulphur as determined by titration = 87.0 ppm

$$\text{Corrected sulphur} = 87.0 \times \frac{0.025}{0.02} = 108.8 \text{ ppm}$$

• Note

The free sulphur dioxide level is measured using exactly the same method described above, but missing out step 2, i.e. by going direct from step 1 to step 3. It is expressed as 'ppm free sulphur dioxide'. The bound sulphur dioxide is the difference between the free and total figures.

The Aspiration method

• Reagents

0.3 per cent hydrogen peroxide Take 10 ml of '100 volume' hydrogen peroxide and dilute to 1 litre with water. This is stable for at least three months at winery temperatures.

25 per cent phosphoric acid Dilute 300 ml of 85 per cent orthophosphoric acid to 1 litre with water.

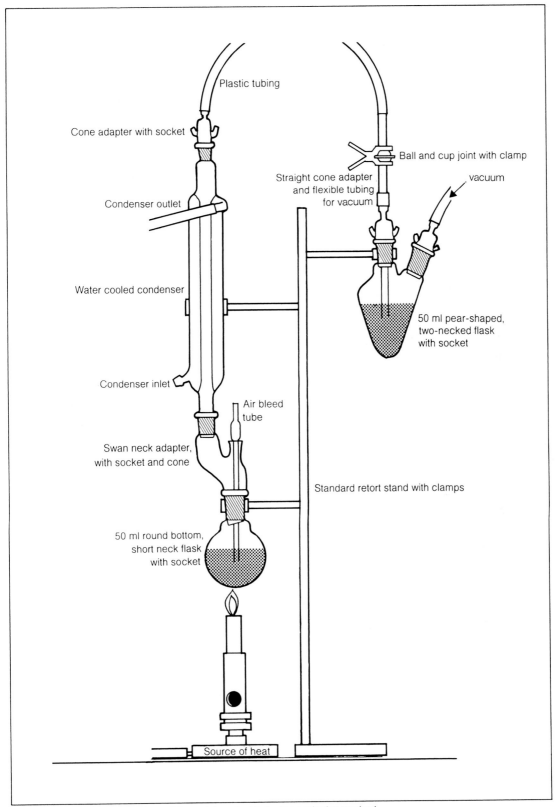

Figure A.4.1 Apparatus for determining sulphur dioxide by the aspiration method

Indicator solution Add 0.1 g methyl-red to 50 ml of 96 per cent aqueous ethanol. Make up to 100 ml with water.

0.01N sodium hydroxide Dissolve 0.4 g sodium hydroxide in 500 ml of water and dilute with water to 1 litre. Alternatively dilute 100ml 0.1N NaOH to 1 litre with water. Store in plastic bottle no longer than seven days.

The laboratory apparatus used in the Aspiration method is shown in Figure A.4.1.

• Method A. 'Free' sulphur dioxide

1. Pipette 10 ml of 0.3 per cent hydrogen peroxide into the two-necked pear-shaped flask and add three drops of the indicator solution. Titrate to the first traces of yellow that persists with the 0.01N sodium hydroxide and connect this flask to the top of the aspiration apparatus. The amount of sodium hydroxide used for colour adjustment does not have to be recorded.

2. Pipette accurately 20 ml of sample into the round-bottomed flask and add 10 ml of 25 per cent phosphoric acid. Attach this flask to the bottom of the apparatus and draw air through the sample for 10 minutes. The colour of the peroxide solution changes to red indicating that sulphur dioxide has been absorbed.

3. Remove the top pear-shaped flask with bubbler still attached and carefully rinse the bubbler tube inside and out with water, the washings remain in the flask. Titrate with 0.01N sodium hydroxide to a yellow end-point. Record the number of mls used in this titration.

4. To calculate the amount of sulphur dioxide in the sample, multiply the number of mls of sodium hydroxide used by 16. Express as 'ppm free sulphur dioxide'.

• Method B. 'Bound' sulphur dioxide

1. After determining the 'free' sulphur dioxide, replace the pear-shaped flask, including the contents and bubbler, to its position on the apparatus. Draw air (aspirate) through the apparatus while boiling the sample for 10 minutes.

2. Remove, once more, the top flask, stop the heating and titrate contents with 0.01 N sodium hydroxide from the burette as before. Record the number of mls used.

3. To calculate the amount of bound sulphur dioxide in the sample, multiply the number of mls used by 16. Record as 'ppm bound sulphur dioxide'.

• Notes

To calculate the total sulphur dioxide content of the sample add the 'free' and 'bound' values.

It is recommended that the accuracy of the techniques be checked regularly, using a known standard of sulphur dioxide. In red wines some sulphur dioxide will be bound to the colouring matter, yet it will be recorded as 'free sulphur dioxide'. Thus the correct figure will be slightly lower. This applies, of course, to the free levels in both methods.

Appendix 5
Determination of the alcohol content in wine

Ebullioscope

This method is based on the variation between the boiling point of water and solutions containing alcohol, and requires the purchase of a special piece of apparatus known as the ebullioscope or ebulliometer. The Dujardin-Salleron type is best known.

- **Method**

1. Rinse and drain the boiling chamber of the ebullioscope with water. Remove the thermometer from the chamber, pipette 25 ml of water in and replace the thermometer. Screw the condenser into position and apply heat.
2. Bring the water in the boiling chamber to the boil and continue heating until the temperature remains constant. Set the slide rule so that the 0.00 mark of alcohol is opposite this temperature.
3. Empty the ebullioscope chamber, rinse with sample and pipette 50 ml of the sample into the chamber. Re-insert the thermometer and screw on the condenser filled with cold water.
4. Apply heat and boil the sample until the thermometer remains static. Note the temperature level, switch off the heat source. In wine samples read off the percentage of alcohol opposite the observed temperature on the scale marked 'wine'.

- **Notes**

Samples of sweet wines should be diluted to a 5 per cent sugar concentration before boiling; the result is multiplied by the dilution factor.

Distillation method

Set up the distilling apparatus as shown in Figure A.5.1.

- **Method**

1. Measure 250 ml of wine sample, using a measuring cylinder, and transfer into the l-litre boiling flask containing about 300 ml of water and a few boiling chips. Connect the condenser to water and apply heat.
2. Distil about 200 ml of the sample into a 250 ml Erlenmeyer flask containing about 20 ml of water, making sure the outlet of the condenser is submerged throughout the distillation.
3. Make up the distillate to 250 ml with water and mix well. Float an alcohol or specific gravity hydrometer in the distillate and record the reading. Make sure the temperature of the sample is identical to that at which the hydrometer is calibrated. The temperature can be adjusted in a water bath if necessary.
4. Use the conversion table (shown in Table A.5.1) if the reading is made with a hydrometer designed for measuring specific gravity.

- **Notes**

Greater accuracy can be achieved by using a pycnometer—see Appendix 1 —to measure the specific gravity in step 3. For an approximate check of alcohol level in wine a 'vinometer' may be used. This is an open-ended glass capillary tube with a tulip-shape opening

Table A.5.1 The relationship between specific gravity and the percentage of alcohol by volume

Specific gravity (at 15°C)	Per cent (v/v alcohol)
1.0000	0.0
0.9992	0.5
0.9985	1.0
0.9978	1.5
0.9970	2.0
0.9963	2.5
0.9956	3.0
0.9949	3.5
0.9972	4.0
0.9935	4.5
0.9928	5.0
0.9922	5.5
0.9915	6.0
0.9909	6.5
0.9902	7.0
0.9896	7.5
0.9890	8.0
0.9884	8.5
0.9878	9.0
0.9872	9.5
0.9866	10.0
0.9860	10.5
0.9855	11.0
0.9849	11.5
0.9844	12.0

on top. A wine sample is poured into this opening and the vinometer is then inverted. The percentage of alcohol is read directly off the scale. Water is used in the same way to check accuracy.

Combi Test

This is an accurate method which can also be used to measure sugar. It is too detailed to describe here, but it is suggested that readers wishing for further information should use the reference given at the end of Appendix 1.

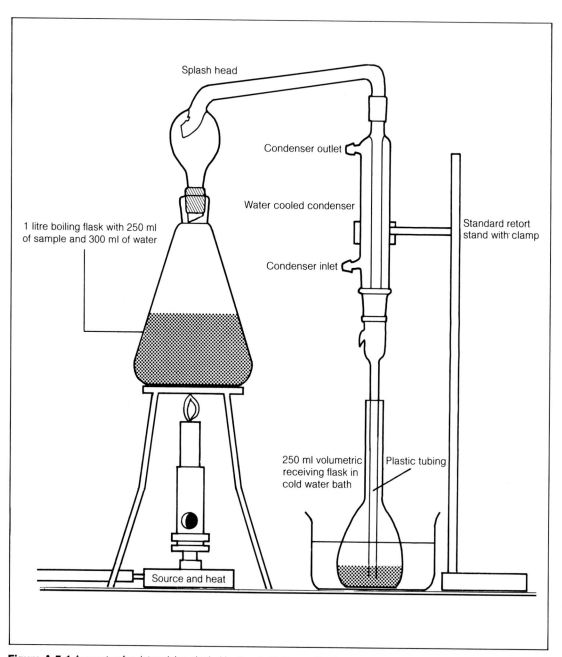

Figure A.5.1 Apparatus for determining alcohol by the distillation method

Appendix 6
Determination of volatile acidity

High volatile acidity indicates spoilage and has an objectionable effect on the wine bouquet and flavour. In most wine-producing countries a legal limit is set for the amount permitted in a wine, and the determination of volatile acidity is an important part of the quality control checks prior to a wine being released on the market. *Small* amounts of acetic and other volatile acids are by-products of a healthy fermentation and, especially with the red wines, they form an important part of the wine's character.

- **Reagents**

0.01N sodium hydroxide Dissolve 0.4 g sodium hydroxide pellets in 500 ml of water and dilute to 1 litre with water. Check normality as described in Appendix 2.

0.3% hydrogen peroxide Dilute 10ml of 30% hydrogen peroxide to 1 litre with water.

Apparatus

The distillation apparatus required for this method is shown in Figure A.6.1.

- **Method**

1. Place 300 ml of water and a few boiling chips in a 1 litre Erlenmeyer boiling flask (A) and connect to the condenser (C). Apply heat to the boiling flask.

2. Place a 250 ml Erlenmeyer flask (flask D) containing 10 ml of water under the outlet of the condenser.

3. Pipette 10 ml sample into the inlet of the Markham apparatus (B) keeping clip (e) open and (f) closed. Add 1 ml of hydrogen peroxide solution and rinse the sides of stopper with distilled water. Replace stopper and fill stopper funnel with water.

4. When water boils open (f) and close (e). Turn on cooling water in condenser (C).

5. Slowly close clamp (g) and steam until 100 ml of condensate is collected in receiving flask (D).

6. Remove receiving flask, close (f) and open (e). Remove stopper and rinse inner wall. open (f) and (g).

7. Repeat previous steps (3-6) using 5 ml distilled water in place of wine sample.

8. Add five drops of phenolphthalein indicator to the distillate in flask (D) and titrate with 0.01N sodium hydroxide until a definite pink colour persists for 15 seconds. Record the number of mls of sodium hydroxide titrated for the sample ('titre sample') and the number for the blank ('titre blank').

9. Volatile acidity in grams acetic acid per litre = (titre sample—titre blank) x 0.06.

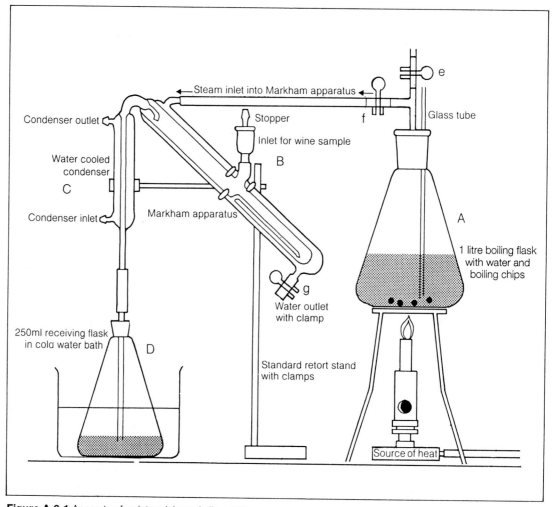

Figure A.6.1 Apparatus for determining volatile acidity

Appendix 7
The detection of malo-lactic fermentation

The conversion of malic acid to lactic acid in red wines, especially in cool-climate wine regions, can be of considerable importance. As conditions favourable to this conversion are also favourable to various undesirable bacterial spoilages, it is essential that the winemaker can measure the malic acid content in his wine. A paper chromatography method is used to determine the presence of malic acid and to check whether malo-lactic fermentation has occurred.

- **Reagents and equipment**

Eluting solvent In a 250ml container add 100ml n-butyl acetate, 40ml formic acid and 10ml distilled water. A measuring cylinder is sufficiently accurate. To this solution add and dissolve 0.075g sodium formate and 0.03g bromo phenol blue indicator.

Malic acid solution Prepare a 5 per cent malic acid solution by dissolving 5 g of L-malic acid in 50 ml of water and make up the total content to 100 ml with water. This will keep for up to four weeks in a refrigerator.

Chromatograph paper Use sheets of 25 cm x 25 cm Whatman No. 1 paper.

- **Method**

1. Make a pencil line about 2 cm from the bottom of the chromatography sheet and mark the line at 2 cm intervals. Place a small drop of the malic acid solution on the first marked point using a capillary tube or a micro-loop made of thin wire. The spot made should be as small as possible, preferably not more than half a centimetre across. Avoid touching the paper with fingers.
2. Place drops of the sample wines along the marked line at 2 cm intervals until all samples are represented. Avoid contamination of the samples by using the correct size of spots and by cleaning and drying the capillary tube or loop with water and tissue paper between successive samples.

3. Allow the spots to dry for minimum of 15 minutes. Then place the chromatography paper in a 2-litre beaker containing no more than a 1 cm depth of solvent. The pencil line and spots must be at the base and above the level of the solvent. To prevent the paper from touching the sides of the beaker, the paper can be curled into a cylinder by clipping the sides together. Cover the beaker with a sheet of glass.
4. Allow to stand for approximately 2-3 hours or until the solvent travels at least 15-20 cm up the paper. Remove the paper and hang it up to dry.

Organic compounds in each spot move upwards, in a column-like fashion, at different rates. The position of acids is indicated by yellow-coloured spots on the dried paper. The malic acid standard, placed as the first drop, will show the level to which malic acid has moved, and yellow spots at similar levels above other samples indicate the presence or absence of this acid. The colour intensity indicates the amount of acid present. The blue-green background of the soaked section of the paper will allow easy detection of the acid spots for the first hour or two after drying. This colour will fade with time, and can be revived by holding the paper in a horizontal position over the open lid of a formic acid bottle. Progress in the conversion of malic to lactic acid can be determined by repeating the test weekly for a month after the cessation of primary fermentation.

The intensity of the malic acid spot from a wine sample will decrease if malo-lactic conversion is occurring. When it disappears, or stays at a constant low intensity, secondary fermentation is complete. In wine samples, acids will appear in the following order from the base line: tartaric, malic, lactic and citric.

Appendix 8
Determination of extract in wine

Wine extract consists of non-volatile components, such as sugars, acids, and salts; these are also known as the 'total soluble solids'. Extract is considered important in European wines, where the finest wines have a high level.

Method

Determine the alcohol content of the wine using one of the methods described in Appendix 5. Using Table A.1.4. establish the specific gravity of an aqueous ethanol solution with the same ethanol concentration (SGe).

Establish the specific gravity of the wine using a hydrometer (SGa). If a Brix hydrometer is used, convert to specific gravity using Table A.1.4.

Specific gravity of the wine (SGw) excluding the alcohol content is given by the following formula: SGw = SGa + 1 - SGe.

To find the extract level use the following conversion:

SGw	Extract (g/l)	SGw	Extract (g/l)
1.000	0.0	1.013	33.6
1.001	2.6	1.014	36.2
1.002	5.1	1.015	38.8
1.003	7.7	1.016	41.3
1.004	10.3	1.017	43.9
1.005	12.9	1.018	46.5
1.006	15.4	1.019	49.1
1.007	18.0	1.020	51.7
1.008	20.6	1.021	54.3
1.009	23.2	1.022	56.9
1.010	25.8	1.023	59.5
1.011	28.4	1.024	62.1
1.012	31.0	1.025	64.7

Sugar-free extract is the extract figure from this table minus residual sugar (Appendix 1). Most wines vary from 0.7 g/100 ml for light wines to 3.0 g/100 ml for heavy wines.

Appendix 9
Determination of chemical stability in wine

Prior to bottling, a wine should be checked for its basic stability. A stable wine should not be affected by moderate temperature change, nor should it vary unduly in chemical composition over time, or be 'micro-biologically unsound'. An unstable wine will need to be treated by one of the methods described in this or the following Appendix.

Methods

Equal quantities of a filtered sample of wine are poured into two sample bottles (A and B). Bottle A is kept for 48 hours at 40°C and bottle B for the same length of time at -2°C.

A wine that remains clear in both bottles will be chemically stable and may then be tested for microbiological stability. One that is not clear will have sediments or hazes of the following types:

1. Crystals. The formation of crystals which are off-white and tasteless, indicates a lack of cold stability. Restabilise wine by chilling at -4°C for minimum of four days and filter prior to bottling.

2. Hazes. The formation of haze in the sample will indicate either chemical or microbiological instability. Chemical instabilities could be metal, phenol, or protein hazes. These can be determined as follows:

Metal hazes* Add 5 ml of concentrated hydrochloric acid to 45 ml of the wine sample; if the haze disappears it was due to the presence of copper or iron.

Phenol hazes A fine sediment or haze can be caused by tannin, or phenol-based pigments. This can be tested by mixing equal quantities of the wine and absolute alcohol; if the haze disappears phenol will be responsible. Gelatine fining followed by filtration will overcome this problem.

Protein hazes This is the haze which will appear during the heat test described above. An alternative method for determining such hazes is also available—the Bento-test**.

Protein hazes are removed by fining with bentonite or similar agents, followed by filtration.

Protein hazes may be due to dead or living yeast cells and to check the presence of the latter the methods described in Appendix 10 are required.

---

* Detailed methods of treating this condition are described by Amerine and Joslyn. 1970. *Table Wines: The Technology of their Production.* University of California Press, Berkeley.

** If not locally available this process, with instructions, may be obtained from C. H. Boehringer Sohn, 6507 Ingelheim am Rhein, West Germany.

Appendix 10
Determination of microbial stability in wine

The activity of yeasts or bacteria in bottled wine can cause serious problems. The correct use of filtration and sterile bottling (see Chapter 11) should prevent this, but the purity checks described here will prove a useful precaution, especially with freshly-bottled sweet wines or red wines which have been prevented from undergoing malo-lactic conversion. Two methods are described.

Agar plating

• Reagents and equipment

Agar 1 per cent Make a 1 per cent solution of potato-dextrose agar (P.D.A. agar) in water. Store in bottles.

Autoclave Or a large pressure-cooker.

Petri dishes

Microscope A standard low-magnification microscope (40-100x) is suitable.

Micro-loop A loop of nichrome wire (e.g. wire from a heater coil) inserted in a wooden handle.

Source of flame Bunsen burner, alcohol lamp or candle.

• Method

1. Autoclave Petri dishes and the 1 per cent agar in bottles (loosen lid) for 15 minutes at a pressure of 15 psi. Pour the liquid agar onto the plates while hot, replace lid, and leave to set.
2. Sterilise the loop of wire in an open flame and transfer a small amount of the sample by streaking across the surface of the agar.
3. Place the sealed plate in an incubator or sealed cupboard at about 25-30°C and keep for 48 hours.
4. Note presence of any type of growth on the agar surface. A microscopic examination will assist in the identificaton of the micro-organisms present—some expert help may be needed. The samples from a correctly-filtered and bot-

tled wine should be clean, giving plates which show no signs of growth.

Plate count technique

The advantage of using this method is that it not only checks purity but also estimates the degree of contamination.

• Method

1. Make up dilutions of each sample, as shown below, and add 1 ml of each to separate and sterile petri dishes.
2. Pour a further amount of still-warm, not hot, sterilised agar medium onto each plate and mix by swirling the Petri dish until the agar sets.
3. Place for 48 hours in an incubator at 25-30°C and observe for the appearance of growth.
4. The number of colonies counted in the plates of contaminated samples is multipled by the dilution factor. This equals the quantity of yeast or bacteria in 1 ml of sample.

• Notes

The following dilutions are used:
(a) 1 ml of sample into 99 ml of salt/water solution. (1:100 ratio).
(b) 1 ml of diluted sample (a), into 99 ml of salt/water solution. (1:10 000 ratio).
(c) 1 ml of diluted sample (b) into 99 ml of salt/water solution. (1:1 000 000 ratio).
To make a salt solution add salt at 5 per cent to tap water and boil for ten minutes. Distilled or chlorinated water should not be used.

With both methods store all the plates upside-down after pouring agar, this will prevent condensation accumulating on the lid and allow easier observation of the agar surface.

A winemaker whose wines show consistent instability, and the presence of viable yeasts and bacteria after bottling, is advised to seek assistance from specialised laboratories in his region. The correct identification of the micro-organisms causing the problem—and future prevention—will then be possible.

NOTES AND REFERENCES

Amerine, M.A., Berg, H.W., Creuss, W.V. 1972. *The Technology of Winemaking*, 3rd ed. Avi Publishing, Westport, Conn.

Amerine, M.A., Berg, H.W., Kunkee, R.E., Ough, C.S., Singleton, V.L., Webb, A.D. 1979. *The Technology of Winemaking*, 4th ed. Avi Publishing, Westport.

Amerine, M.A., Joslyn, M.A. 1970. *Table Wines: The Technology of Their Production*, 2nd ed. Univ. of Calif. Press, Berkeley and Los Angeles.

Amerine, M.A., Ough, C.S. 1980. *Methods for the Analysis of Musts and Wines*, John Wiley, New York.

Amerine, M.A., Roessler, E.B. 1976. *Wines, Their Sensory Evaluation*, W.H. Freeman, San Francisco.

Amerine, M.A., Singleton, V.L. 1977. *Wine, An Introduction for Americans*, 2nd ed. Univ. California Press, Berkeley.

A.O.A.C. 1965. *The Official Methods of Analyses*, 10th ed. Association of Agric. Chemists Press, Washington.

Becker, H., Kerridge, G.H. 1972. Methods of small-scale winemaking for research purposes in both hot and cool regions. *Journal Australian Institute Agricultural Science* 38:3-6.

Broadbent, J.M. 1973. *Wine Tasting, A Practical Hand book on Tastings*, Christie Wine Publications London.

Eschenbruch, R., Sage, N.F. 1976. Small-scale winemaking at Te Kauwhata viticultural Station, *Food Technology in N.Z.*, 11: 15-19.

Geiss, W. 1960. *Lehrbuch für Wein Bereitung und Kellerwirtschaft*, Bad Kreuznach.

Hennig, K., Jacob, L. 1973. *Untersuchungsmethoden für Wein und Ahnliche Getranke*, Verlag Eugen Ulmer, Stuttgart.

Iland, P. G. 1988. *Techniques for Accurate Chemical Analysis of Grape Juice and Wine.* Patrick Iland Publ. S. Aust.

Jackisch, P. 1985. *Modern Winemaking*, Cornell Univ. Press, Ithaca.

Jeffs, J. 1971. *The Wines of Europe*, Faber & Faber, London.

Kramer, O. 1954. *Kellerwirtschaflisches Lexicon*, Verlag-Meininger Press, Neustadt.

Ough, C.S. 1992. *Winemaking Basics*, Food Products Press, New York.

Peynaud, E. 1984. *Knowing and Making Wine* (English translation A. Spencer), Wiley Publishers, New York.

Rankine B. 1989. *Making Good Wine*, Winetitles, Adelaide.

Ribereau-Gayon, J., Peynaud, E. 1961. *Traite d'Oenologie*, Beranger, Paris.

Roupnel, G. 1911. *La Bourgogne*, Horizons of France Press, Paris.

Ruakura Agr. Research Centre, Hamilton, *N.Z. Oenological & Viticultural Bulletins* 1-11, 13-15, 17, 19-21, 24, 26, 27.

Tchelistcheff, A., Peterson, R.G. and van Gelderen, M. 1971. Control of malo-lactic fermentation in wine, *American Journal of Enology & Viticulture*, 22:1-5.

Troost, G. 1972. *Technologie des Weines*, Verlag Eugen Ulmer, Stuttgart.

Vine, R.P. 1981. *Commercial Winemaking: Processing and Controls*, Avi, Westport, Conn.

Webb, A.D. (Editor) 1974. *Chemistry of Winemaking*, Advances in Chemistry Series 137—American Chemical Society, Washington, D.C.

White R., Adamson, B., and Rankine, B. 1989. *Refrigeration for winemakers*, Winetitles, Adelaide.

INDEX